Impact Studies

Within the past two decades, research on impact cratering mainly gained attention throughout the world after the theory that a large impact event caused the extinction of ca. 50% of all living species, including the dinosaurs, approximately 65 million years ago was proposed. Impact craters are formed when a large meteorid crashes into a larger planetary body with a solid surface.

The interdisciplinary series, **Impact Studies**, aims to include all aspects of research related to impact cratering - geology, geophysics, paleontology, geochemistry, mineralogy, petrology, planetology, etc. Experimental studies on shock metamorphism, investigations of individual impact craters, comparative planetology, and paleontological research on the effects of impact events are just a few topics that are treated in this series. All contributions are peer-reviewed to ensure high scientific quality.

More information about this series at http://www.springer.com/series/4698

Victor L. Masaitis · Mikhail V. Naumov
Editors

The Puchezh-Katunki Impact Crater

Geology and Origin

Springer

Editors
Victor L. Masaitis
A. P. Karpinsky All-Russia Institute of Geological Research (VSEGEI)
St. Petersburg, Russia

Mikhail V. Naumov
A. P. Karpinsky All-Russia Institute of Geological Research (VSEGEI)
St. Petersburg, Russia

ISSN 1612-8338
Impact Studies
ISBN 978-3-030-32042-3 ISBN 978-3-030-32043-0 (eBook)
https://doi.org/10.1007/978-3-030-32043-0

Cover illustration: Dona Jalufka, Vienna, Austria

This Springer imprint is published by the registered company Springer Nature Switzerland AG
The registered company address is: Gewerbestrasse 11, 6330 Cham, Switzerland

Preface

The discovery and geological study of ancient impact structures (or astroblemes) on Russia's territory was first started about half a century ago. For a long time, many of such structures were considered as a result of ordinary geologic processes—volcanism, tectonic movements, erosion, etc. Detailed studies of rocks from these structures showed that they were subjected to high-pressure transformations, which may be put into effect only due to high-velocity collisions of small cosmic bodies falling on the Earth. The studies of terrestrial impact structures acquired a wide scope throughout the world, and these studies coincided with the beginning of the era of space research using automatic and manned spacecrafts. The investigation of surfaces of solid planets and various asteroids showed that processes of impact cratering play a fundamental role in the Solar System. The important influence of these processes on the formation of the early Earth's crust and formation of various mineral raw materials as well as on biotic changes, etc. was revealed during geological studies of Earth's land. About two hundreds terrestrial impact structures have already been found, their number increasing every year not only due to new discoveries in geologically poorly studied areas but also as a result of the use of new diagnostic methods in previously well-studied regions.

The interest in the impact structures and in their rock composition, in addition to the reasons noted above, gives rise by determining of effects of cratering in various terrestrial paleoenvironments, by the reconstruction of collision parameters, and projectiles composition. All this can be used for better understanding of the processes of impact cratering on different planets and asteroids and history of their surfaces.

To date, about 20 ancient impact structures including Popigai (100 km in diameter), Puchezh-Katunki (80 km) and Kara (65 km) are discovered in Russia. All of them, especially their interiors and composition of crater's fill, have undergone detailed geological studies in one way or another. The obtained data have been given in many publications in Russian, but only limited of them were translated into English. An exception is the book "Popigai impact structure and its diamond-bearing rocks" published by Springer in 2019. The present book represents in detail the Puchezh-Katunki impact structure, its interior features, rock composition and mode of origin and makes it possible for English-speaking readers to get acquainted with results of detailed studies of this large crater.

Puchezh-Katunki impact structure is situated in central Russia at about 400 km northeast of Moscow. For a long time, from the middle of the nineteenth century up to the 1960–1970s, the origin of various dislocations in the site of Puchezh-Katunki has been debated. The first statement on the possible extraterrestrial origin of Puchezh-Katunki dislocations and host structure was advanced only in 1965. Some years after, it was confirmed by a detailed study of rocks transformed by shock metamorphism. The comprehensive investigation program of Puchezh-Katunki was carried out in 1988–1992 in connection with the drilling in the center of the crater of the Vorotilovo Deep Well (VDW), which reached 5374 m deep. The deep drilling in the center of a large impact structure was performed for the first time and enabled to obtain important information on rock transformations occurred in the deep-seated zones beneath the crater. In frames of the investigation program, versatile research works were performed by geologists of JSC "SIC Nedra" (Yaroslavl), VSEGEI (St. Petersburg) and some other scientific institutions. They include, in addition to logging and detailed core study, a comprehensive analysis of materials obtained from previous drilling (including core logging of tens of earlier boreholes), geological mapping and geophysical surveys as well as the dynamic modeling of origin of large impact craters. The results of this project were summarized in the book "Deep Drilling in the Puchezh-Katunki impact structure" published in Russian by VSEGEI Press in 1999 under edition of V. L. Masaitis and L. A. Pevzner.

Some additional mineralogical, petrographic, geochemical, palynological and isotope studies of rock samples from Puchezh-Katunki were carried out in recent decades providing a lot of new important information on impact-derived and impact-transformed formations as well as on the age of the impact event. Some of modern concepts of impact cratering on Earth and other planets, of the composition, occurrence and transformation of various rocks required additional changes of previously obtained results. The numerical simulation based on modern programs gives an opportunity to improve the impact cratering model for Puchezh-Katunki. Besides, a significant development of the International Stratigraphic Chart required a corresponding improvement of stratigraphic terminology.

Thus, the present book provides up-to-date knowledge on Puchezh-Katunki impact structure, on its interior and impact lithologies. It includes two parts. In the first part, the geological framework of the impact structure, impact-generated rock formations and effects of impact-derived rock transformations are described. The second part concerns the model of impact cratering and processes of origin of Puchezh-Katunki structure. The book is compiled by an author's team, which took a principal part in the investigation program mentioned above. It is written by V. Masaitis, M. Mashchak, M. Naumov, T. Selivanovskaya, and J. Orlova from A. P. Karpinsky Russian Geological Research Institute (VSEGEI), and B. Ivanov from Institute of Dynamics of Geospheres of Russian Academy of Science. The book is edited by V. L. Masaitis and M. V. Naumov (VSEGEI).

We acknowledge Dr. Alex Deutsch (Institute of Planetology, University of Münster, Germany) acquainted with some parts of the primary text and made remarks and advice, which were taken into consideration. This made it possible to improve considerably some sections and eliminate inaccuracies.

St. Petersburg, Russia

Victor L. Masaitis
Mikhail V. Naumov

Contents

Abbreviations

Bh	Borehole
DTA	Differential thermal analysis
EMP	Electron microprobe
ICP-MS	Inductively coupled plasma mass spectrometry
ICP-OES	Inductively coupled plasma optical emission spectrometry
LOI	Loss of ignition
RAS	Russian Academy of Science
SEM	Scanning electron microscopy
TEM	Transmission electron microscopy
VDW	Vorotilovo Deep Well
VSEGEI	A. P. Karpinsky Russian Geological Research Institute
XRF	X-ray fluorescence analysis

Abbreviation of Mineral Species (According to Recommendations of the International Mineralogical Association (IMA))

Ab	Albite
An	Anorthite
Bt	Biotite
Cord	Cordierite
CPx	Clinopyroxene
En	Enstatite
Fs	Ferrosilite
Fsp	Feldspar
Grt	Garnet
Gs	Impact glass
Mt	Magnetite
OPx	Orthopyroxene

Ort	Orthoclase
Pl	Plagioclase
Py	Pyroxene
Sill	Sillimanite
Woll	Wollastonite

Geological Framework of the Puchezh-Katunki Impact Structure

Research History of the Disturbances in the Center of East European Platform

Mikhail S. Mashchak and Mikhail V. Naumov

The peculiar dislocations of Triassic and Permian sedimentary rocks and associated variegated breccias exposed on the banks of the Volga River between settlements Sokolskoye and Gorodets for more than 50 km, as well as the isolated protrusion of Early Precambrian crystalline basement rocks in the vicinity of the village of Tonkovo (about 30 km east of the Volga River), were a geological mystery for a long time. Several attempts have been undertaken to understand the genesis of breccias and dislocations. Detailed reviews of early studies performed and of different views on the origin and age of the dislocations were given by Schvetsov (1939), Kudinova (1939), Bakirov (1948), Kazakov (1950), Nechitailo et al. (1959) and Goretsky (1962). All the proposed hypotheses were based on only restricted observations and usually accounted for certain geological features of studied objects. Therefore, they were disproved by subsequent studies.

The first geognostic information on the Puchezh-Katunki dislocations goes back to the middle of the XIX century when Roderick I. Murchison had described in detail variegated breccias outcropped along the banks of the Volga River (Fig. 1) in his geological description of the Volga River vicinities between towns of Yurievets and Balakhna (Murchison 1845). He suggested the origin of the breccias by metamorphism of the underlying Permian limestone. Later, these breccias were considered as ancient landslides (Nikitin 1885), or ancient karst feature (Amalitsky 1887). There were a number of viewpoints on the origin of the dislocations in connection with local vertical tectonic movements (e.g., Zemyatchensky 1886; Sibirtsev 1896; Noinsky 1932; Moldavskaya 1933).

During 1930ths–1940ths, the area of the Puchezh-Katunki dislocations and associated variegated breccias were studied in detail (including shallow drilling and electrical methods) due to exploration for the projected Gorky reservoir. In 1933, the so-called "conglomerate-breccia-clayey strata" had been distinguished as a specific

M. S. Mashchak · M. V. Naumov (✉)
A.P. Karpinsky Russian Geological Research Institute,
199106 Sredny Prospekt 74, St. Petersburg, Russia
e-mail: m_naumov@mail.ru; Mikhail_Naumov@vsegei.ru

V. L Masaitis and M. V Naumov (eds.), *The Puchezh-Katunki Impact Crater*,
Impact Studies, https://doi.org/10.1007/978-3-030-32043-0_1

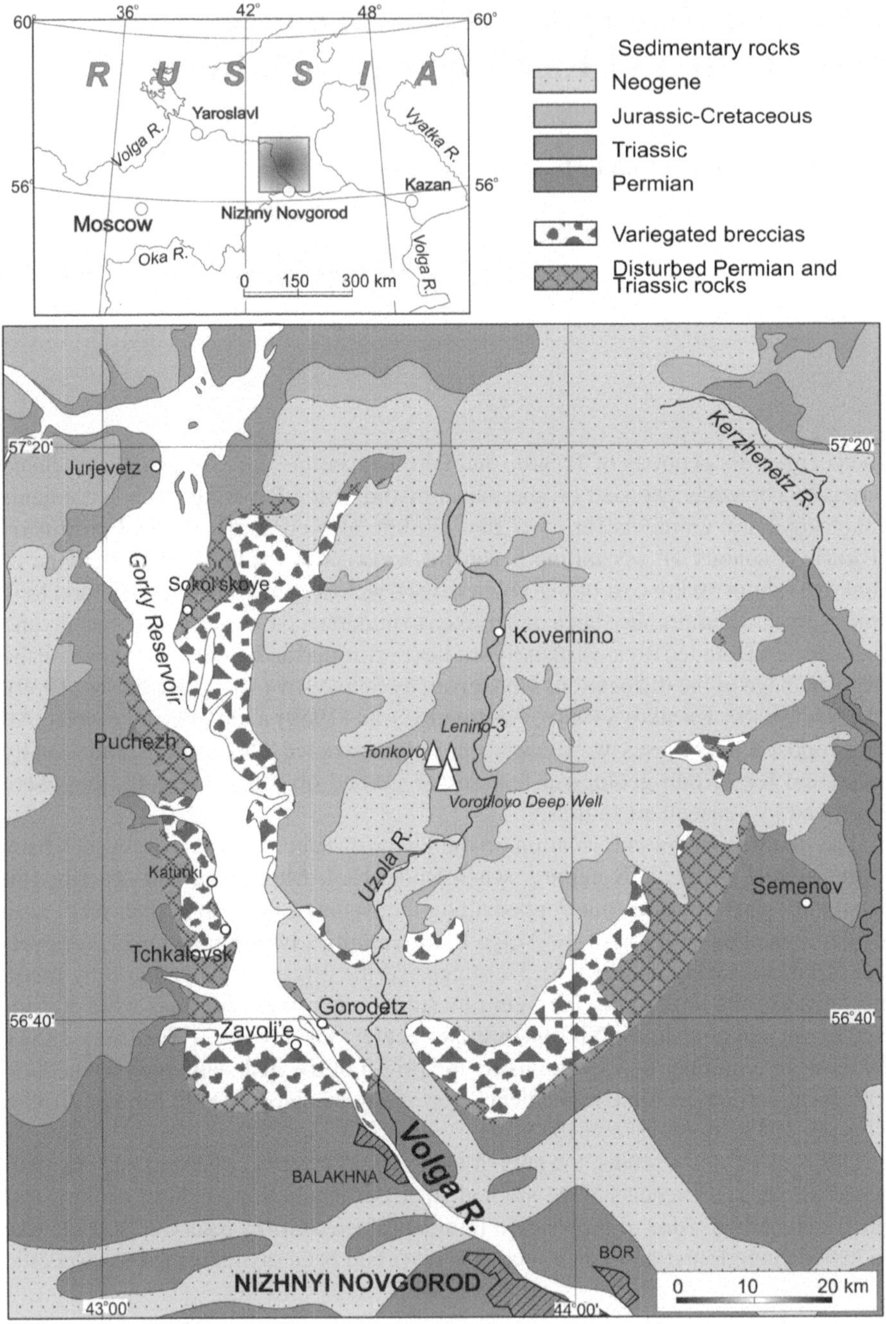

Fig. 1 Geological Map of the Gorodets-Kovernino area (simplified from Fadeeva 2000a). Quaternary deposits are removed. The map shows the surface geology of the area where Puchezh-Katuki impact structure is situated. In addition, location of so-called Puchezh-Katunki dislocations is demonstrated. Are shown boreholes mentioned in this chapter

formation (Kudinova 1939). In particular, a near-surface character for the dislocations of Permian to Triassic sedimentary rocks had been revealed and the Triassic age for the breccias had been proposed. As a result, a tectonic origin of the dislocations was discarded, and a number of new ideas appeared, namely: Quaternary glacial dislocations (e.g., Kudinova 1939), Triassic glacial dislocations (Ivanov 1949; Tikhvinskaya 1956), ancient salt tectonics (e.g., Shvetsov 1939), mudflows and proluvial flows in the Triassic (Bakirov 1948; Kazakov 1950).

The latter idea was processed by Shatsky (1946) to the "hypothesis of gravitational tectonogenesis". According to this hypothesis, the dislocations resulted from slipping of rock masses occurred on uplift slopes due to both oscillatory crustal movements and erosion processes that occurred in the Triassic and, probably, in the Early Jurassic at a stage of sedimentation and early diagenesis. In this conception, the gravity has been assumed to be the principal cause for dislocations. Episodic powerful showers caused the intense drift of loose weathered material downwards the slopes in areas where clayey and marl deposits occur under arid environments.

Another view on the origin of the dislocations was the hypothesis of injective tectonics (Goretsky 1962). This author considered that the deformations had been caused mostly by radial tectonic movements during the Triassic. These deformations have provided irregular strains, which influenced on plastic and brittle lithologies by different ways. Intense lateral pressure caused the creation of small surface brachy structures detaching of plastic sediments, formation of local dislocation centers and injections of older deposits into the overlying younger strata. As to Precambrian gneisses opened by drilling in the vicinity of Tonkovo village, Goretsky (1962) assumed that those rocks are blocks and clumps injected to Paleozoic sediments but not the crystalline bedrock.

Incidentally, many above researchers (e.g., Kazakov 1950; Goretsky 1962) paid attention to the similarity of the dislocations on the Volga banks with disturbances at some areas where impact structures were proven later (e.g., Karla, midstream of Volga River, Russia, or Wells Creek, Tennessee, USA).

A sharp change in views on the origin of dislocations happened in 1950ths when a protrusion of Early Precambrian crystalline basement had been discovered at about 30–40 km east of the Puchezh-Katunki dislocation zone. As early as in 1940ths, a vast roughly circular area of about 80 km across east of the Volga River was paid a special attention due to young (Cretaceous to Tertiary) sedimentary formations developing there. As a result, an annular depression of about 42 km in diameter filled in by a 400-m thick clayey formation had been outlined by hydrocarbon exploration drilling during 1948–1952 (Nechitailo et al. 1959). It was named as "Kovernino depression" due to Kovernino settlement in its centre. In addition, this exploration showed that the variegated breccias similar to those from the Puchezh-Katunki dislocations area are widespread just outside from the contour of Kovernino depression. Moreover, another breccia containing clastic material originated from the Precambrian crystalline rocks had been opened by drilling in the central part of Kovernino depression. Thus, an uplift of the crystalline basement was supposed within the depression (Ljutkevitch and Frukht 1954). Such an uplift had been discovered in 1955 when an extensive magnetic anomaly close to Tonkovo settlement was drilled. The uplift was known as

"Vorotilovo protrusion" while the surrounding area of disturbed rocks of 70–90 km across, as "Gorodets-Kovernino tectonic zone" (Nechitailo et al. 1959).

From this time, the majority of involved geologists considered the dislocations in relation with intense tectonic movements, which caused the formation of the uplift of crystalline basement. According to this hypothesis, the sliding of clumps of Paleozoic rocks from slopes of the protrusion should have caused breaking, crushing, and overtilting in the Paleozoic rocks (Nechitailo et al. 1959; Blom 1960; Susalnikova 1965a, and others).

The discovery of the Vorotilovo protrusion stipulated hard regional geophysical studies, drilling and scientific research works including detailed aeromagnetic, gravity, and electrical surveys, and refraction shooting (Nechitailo et al. 1959; Susalnikova 1965b; Tumanov 1973, and others). The results of these investigations gave rise to develop new views on the structure of the area. It was found that the Vorotilovo protrusion is a local uplift of no more than 8–10 km across and of 1.5–2 km high. In addition, enigmatic "volcanic rocks" were opened around the protrusion (Ilyna and Frukht 1963). Therefore, a volcano-tectonic, explosive origin of the dislocations was assumed. For example, Soloviev (1958) considered the "dislocations of variegated deposits" to be the consequence of tectonic processes, that should have occurred from the Middle Cretaceous up to the Quaternary; the intense disruption of the basement was caused by powerful explosions stipulated by the decay of radioactive elements. Vardanyants (1961) suggested that the Vorotilovo protrusion is a diatreme formed due to gas explosions within the crystalline basement.

An idea of the volcanic-tectonic origin of the Puchezh-Katunki dislocations and Vorotilovo protrusion has been elaborated in 1960ths by R. R. Tumanov, who headed the geological mapping of 1:200,000 scale in the region. According to the conception in question, the circular structure including Vorotilovo protrusion, Kovernino depression, and Puchezh-Katunki dislocations, is a multi-stage volcanic-tectonic structure, the formation of which started in the Late Permian and completed in the Middle Jurassic (Tumanov 1973; Tumanov and Bogorodskaya 1975; Bogorodskaya and Tumanov 1980). During the initial stage (Late Permian), a blocky tectonic uplift had arisen. Its sharp elevation in the early Middle Triassic resulted in the formation of ring and radial fault systems and low-amplitude steep folds. The intense erosion, which accompanied the uplifting, was favorable for the accumulation of enormous mass of detrital material, which was transported downslope by mudflows. In the end of the Late Triassic, the Kovernino depression arose; the subsidence was accompanied by the compensation horst-shaped uplift of the crystalline basement in the center of the depression. The sharp elevation of the basement caused its intense fracturing and explosion degassing of the volatiles. Repeated accumulations of gases within separated chambers led to local sub-surface explosions and to intense faulting of the protrusion.

R. R. Tumanov had to suggest a special stratigraphic scheme for the Kovernino area because of formations, which have been found there, neither have analogues in adjacent areas nor elsewhere in East European Platform. In this scheme, five specific units have been singled out (Tumanov 1973; Tumanov and Bogorodskaya

1975). (a) So-called "fanglomerate formation" of presumably Middle to Upper Triassic age that comprises variegated breccias after sedimentary rocks, mostly of Permian and Lower Triassic; later it was entitled as "Puchezh unit" and attached to the Lower Jurassic (unpublished Tech. Rep. of Volgageologija by V.I. Bel'kevich); (b) "Breccia-agglomerate formation" (presumably Lower Jurassic) including megabrecias after crystalline target rocks; (c) sequence of clays and claystones with sandstones and breccias (presumably Lower to Middle Jurassic), (d) Uzola Formation (Middle Jurassic)—various polymict breccia including those containing "volcanic" material. Based on subsequent more detailed investigations, M. N. Kolesov, O. S. Bogatyrev and others (unpublished Technical Reports of Middle Volga Geological Enterprise) subdivided the Uzola Formation into a lower member composed of tuffs, tuff breccias, and mesobreccias, and an upper member consisting of tuffstones, sandstones, and tuffaceous gravelites. In addition, these researchers distinguished the "Emokhony subvolcanic complex" including "basalt-like" bodies and associated "tuffs"; (e) Kovernino Formation (Middle Jurassic)—horizontally-bedded clay sequence filling in Kovernino depression. Except for the latter formation, both lithological, spatial, generic, and age criteria for distinguishing of all listed units were extremely poor.

Meanwhile at the same period, an alternative view on the origin of the "Gorodets-Kovernino tectonic zone" had been suggested. Based on the analysis and comparison of structural data for the above-mentioned zone and world's known impact craters, Firsov (1965) proposed for the first time the idea that the impact of an extraterrestrial body caused the dislocations, whose structure and morphology resembled a gentle funnel of a surface explosion. This impact crater entitled "Puchezh-Katunki" probably formed in the Late Triassic or Early Jurassic (Firsov 1965).

Based on the discovering of shock metamorphic effects in crystalline rocks found in breccia penetrated by the borehole Lenino-3 at the Vorotilovo protrusion, Masaitis (1973) substantiated the meteorite impact origin of the Puchezh-Katunki structure. Hereon, thorough petrographic investigations of cores from numerous boreholes, as well as field studies were performed in the region of the Puchezh-Katunki dislocations by geologists of VSEGEI. During these studies, for the first time the presence of numerous shock features including impact melts was documented. These observations provided the final proof for the extraterrestrial origin of the Puchezh-Katunki structure, which was now interpreted as an impact structure (astrobleme) of Jurassic age (Masaitis 1973, 1974; Masaitis et al. 1980). In this new concept, Vorotilovo protrusion is interpreted as the central uplift of the crater, Kovernino depression, as annular trough, and zone of Puchezh-Katunki dislocations, as western segment of the outer ring terrace. A new subdivision of lithological units was worked out in accordance with the impact concept (Masaitis et al. 1980). So-called "Puchezh unit" is in fact allochtonous megabreccia and klippe breccia after various rocks of the sedimentary platform cover. "Breccia-agglomerate" formation is regarded as shocked, brecciated and cataclased crystalline target rocks developed in the upper part of the central uplift (crystalline authigenic breccia); "brecciated claystones with siltstone interbeds", as brecciated Neoproterozoic and Devonian rocks (sedimentary authigenic breccia). The term "Usola Formation" is reserved only for its upper member, which consists of redeposited in the crater lake material of impact breccias and impactites. Within the lower member, which embraces polymict impact-derived allochtonous breccias, several units are distinguished: polymict microbreccia, polymict allogenic mesobreccia, and suevite. At last, Kovernino Formation represents an upper, thicker portion of crater lake deposits.

During 1970ths and 1980ths, the Gorodets-Kovernino area remained a target for the intense exploration including drilling, large-scale mapping, prospecting, and seismic profiling. Together with previous investigations, this assigns the Puchezh-Katunki as one of the best explored impact structures where impact-generated and shock-metamorphosed formations are opened by 125 deep (>300 m) boreholes, many of them (47) are more than 500 m deep, and 5 holes exceed 1 km deep.

These investigations provided a lot of new information on geological framework, tectonics, lithology, and commodities of Puchezh-Katunki circular structure. Nevertheless, the volcanic-tectonic concept for origin of the latter was reiterated inertially until the end of 1990ths by many local geologists who carried out mapping and prospecting in the region; this concept was presented in Technical Reports of Middle Volga Geological Enterprise prepared by A.A. Menyakin, M.N. Kolesov, O.S. Bogatyrev, A.V. Durandin, V.I. Bel'kevich, and others. All the new data were in full accordance with the extraterrestrial origin of the Puchezh-Katunki. In particular, prospecting works headed by O.S. Bogatyrev, reveal impact diamonds in some impact lithologies. However, the above researchers consider the Vorotilovo protrusion as a multi-ring and multi-phase volcanic structure with the central large funnel and a series (up to 11) of shallow necks (e.g. Durandin 2000; Fadeeva 2000b). At last, this obsolete volcano-tectonic hypothesis has been again discussed by Marakushev et al. (1991, 1993a, b) despite the presence of overwhelming proof for the impact origin. These papers include many invalid assumptions, incorrect interpretations of geological and petrological data, and fantastic hypotheses; thereby, they do not deserve thorough consideration.

In spite of a vast complex of geological and geophysical works and boring held for the last three decades of XX century, the volcanic-tectonic hypothesis has not received additional grounding. On the contrary, a huge additional information on geological framework of the impact crater, composition and distribution of impact-generated formations, shock and post-shock transformations of target rocks, etc. has been obtained in 1988–1992 from the deep drilling accompanied by a comprehensive investigation program.

The Vorotilovo deep well (VDW) (Fig. 2) has been drilled by the Scientific Industrial Center on Super-Deep Drilling and Comprehensive Studies of the Earth ("Nedra") in the center of the Puchezh-Katunki circular structure (Khakhaev et al. 1992). The purpose of this well, which was stopped at 5374 m, was to establish characteristic features of the deep structure of the Puchezh-Katunki crater and to estimate its economic potential (in particular, for use as a deep gas reservoir). A pilot well 100 m north of the VDW site was drilled to a depth of 1498 m. Drilling of both wells was carried out with continuous coring; the total length of recovered cores was 3082 m. Research works included in addition geological and geophysical logging; petrographic, mineralogical, and geochemical study of core; revealing of variation

Fig. 2 Boring tower of the Vorotilovo Deep Well (September 1991)

of physical properties of rocks with depth; checking of the geophysical boundaries; study of fluid regime and sub-surface fluids; etc. All these works were carried out with the participation of VSEGEI geologists who revised a significant part of previously obtained geological data including the core of more than 50 deep boreholes. The deep drilling project as a whole contributes considerably to both recognition of inner structure of the impact crater and its modelling because both lateral and depth (for 5 km) variations of shock, post-shock thermal, and post-impact hydrothermal transformation should be studied (Masaitis and Pevzner 1999).

Data accumulated during deep drilling and analysis of other materials unambiguously testify that the Puchezh-Katunki ring structure is an ancient impact crater ("astrobleme"). This nowadays widely undisputed assessment is based on well-known, clear, and tested criteria for the extraterrestrial origin of a number of circular structures on Earth (Short and French 1968; Masaitis 2005; French and Koeberl 2010; Osinski and Pierazzo 2012; and many others). These features characteristic and relevant for the impact origin of the Puchezh-Katunki structure are published in several papers and books (Masaitis et al. 1980, 1993, 1995, 1996; Feldman et al. 1984, 1992, 1994; Masaitis and Naumov 1990, 1993; Mashchak et al. 1990; Masaitis and Mashchak 1990, 1996; Masaitis 1991; Pevzner et al. 1992; Naumov 1993, 2002; Masaitis and Pevzner 1999). At last, the extraterrestrial (impact) origin of the rock complex composing the Puchezh-Katunki structure was approved in both standard legends for State Geological Maps of Russian Federation (Kirikov 2002; Ulanov et al. 2005) and State Geological Maps themselves (Kochergina et al. 2018; Kirkov et al. 2019).

References

Amalitzky VP (1887) Permian deposits in the Oka-Volga basin. St. Petersburg (in Russian)

Bakirov AA (1948) Geological structure and oil-and-gas contents in Paleozoic structures of the Middle Russian syneclise. Gostoptekhizdat, Moscow (in Russian)

Blom GI (1960) On age of the conglomerate-clay sequence of Kerzhenets, Linda, Vedomost', and Serga rivers basins. Sov Geol 4:118–120 (in Russian)

Bogorodskaya OA, Tumanov RR (1980) Geological map of the USSR on 1:200,000 scale. Sheet O-38-XXVI: explanatory notes. Moscow, 132p (in Russian)

Durandin AV (2000) Geology and diamond-bearing prospects of the Puchezh-Katunki zone. Geol Okhara Nedr 10:28–32 (in Russian)

Fadeeva LI (ed) (2000a) State Geological Map of Russian Federation of 1:000,000 scale (new generation), Sheet O-37, (38)—Nizhny Novgorod. VSEGEI Press, St. Petersburg (in Russian)

Fadeeva LI (ed) (2000b) Explanatory notes to State Geological Map of Russian Federation of 1:000,000 scale (new generation), Sheet O-37, (38)—Nizhny Novgorod. VSEGEI Press, St. Petersburg, p 262 (in Russian)

Feldman VI, Sazonova LV, Nosova AA (1984) Geological structure and petrography of impactites of the Puchezh-Katunki astrobleme (Volga Region). Bull Moscow Soc Nat Res Sect Geol 6:53–63 (in Russian)

Feldman V, Rosen O, Kapustkina I (1992) Tagamites from Puchezh-Katunki astrobleme, Russia (abs.). In: Abstracts for 16th International Russian-American microsymposium on Planetology. Vernadsky Institute, Moscow, pp 29–31

Feldman V, Kotelnikov S, Sazonova L, Guseva E (1994) Diaplectic transformation in clinopyroxene (Puchezh-Katunki astrobleme, Russia) (abs.). In: Lunar Planetary Science, vol XXV, Houston, pp 369–370

Firsov LV (1965) On meteoritic origin of the Puchezh-Katunki crater. Geotectonika 2:106–118 (in Russian)

French BM, Koeberl C (2010) The convincing identification of terrestrial meteorite impact structures: what works, what doesn't and why. Earth Sci Rev 98:123–170

Goretsky GI (1962) On understanding of the nature of Puchezh-Balakhna dislocations (on shows of injective tectonics in the Russian Platform). Bull Moscow Soc Nat Res Sect Geol 5:80–110 (in Russian)

Ilyina NS, Frukht DL (1963) On the problem of volcanogenic rock distribution in the Transvolgian area of the Gorky Region. Trans Dokl Acad Sci USSR 153:906–908 (in Russian)

Ivanov AN (1949) The problem of genesis for the conglomerate-breccia formation of the Puchezh-Gorodetz region. Trans Kazan Univ 109(3) (in Russian)

Kazakov MP (1950) On the nature of Puchezh-Chkalovsk dislocations. Bull Moscow Soc Nat Res Sect Geol 25:28–42 (in Russian)

Khakhaev BN, Masaitis VL, Pevzner LA (1992) Deep scientific drilling in the Puchezh-Katunki impact structure (abs.). In: Proceedings of the 7th international symposium on deep drilling of continental crust, pp 208–210

Kirikov VP (ed) (2002) Legend of the state geological map of Russian Federation of 1:1 000,000 (3d generation). Central European series of sheets. Moscow, SIC "Geocentre" (in Russian)

Kirkov IG, Kochergina VA, Khaidarova DA, Naumov MV (2019) Geological map of Russia Federation on 1:200,000 scale, 2nd generation. sheet O-38-XXVI (Puchezh): explanatory notes. Moscow, 130p (in press) (in Russian)

Kochergina VA, Kirkov IG, Khaidarova DA (2018) Geological map of Russia Federation on 1:200,000 scale, 2nd generation. Sheet O-38-XXVII (Semenov): explanatory notes, Moscow, 90p (in Russian)

Kudinova EA (1939) Geology of Puchez-Chkalovsk area (for criticism of the salt tectonics hypothesis). Bull Moscow Soc Nat Res Sect Geol 17(4–5) (in Russian)

Ljutkevitch EI, Frukht DL (1954) On possible occurrence of a buried ridge of the platform basement in the Gorky region. Trans Dokl Acad Sci USSR 25(2)

Marakushev AA, Bogatyrev OS, Fenogenov AD, Fedosova SV (1991) Endogenic origin for the Puchezh-Katunki ring structure in the Russian platform. Trans Dokl Russ Acad Sci 328:364–368 (in Russian)

Marakushev AA, Budkov GL, Fenogenov AD, Paneyakh NA, Fedosova SV (1993a) Formation of the Puchezh-Katunki ring structure in the Russian platform. Trans Dokl Russ Acad Sci 320:687–691 (in Russian)

Marakushev AA, Bogatyrev OS, Paneyakh NA, Fenogenov AN, Fedosova SV (1993b) Time length of formation of large ring craters as the main argument of their endogenic nature. Geol Okhrana Nedr 4:3–16 (in Russian)

Masaitis VL (1973) Geological consequences of crater-forming meteorite fall. Nedra Press, Leningrad, p 20 (in Russian)

Masaitis VL (1974) Some ancient meteoritic craters in the USSR. Meteoritika 33:64–68 (in Russian)

Masaitis VL (1991) Study of deep interior of impact structures (in Russian). Geol I Okhrana Nedr 8:10–137

Masaitis VL (2005) Morphological, structural and lithological records of terrestrial impacts: an overview. Aust J Earth Sci 52:509–528

Masaitis VL, Mashchak MS (1990) Puchezh-Katunki astrobleme: structure of central uplift and transformation of composed rocks. Meteoritics 4:383

Masaitis VL, Mashchak MS (1996) Recrystallization and blastesis of shock-metamorphosed rocks in impact structures. Zap Vserossiskogo Mineral Obshch 125:1–18 (in Russian)

Masaitis VL, Naumov MV (1990) Structural observations in the west slope of the Puchezh-Katunki astrobleme (abs.). In: Abstracts for 21st all-union meteoritic conference, Moscow, pp 135–136 (in Russian)

Masaitis VL, Naumov MV (1993) Puchezh-Katunki impact crater: preliminary model of hydrothermal circulation system. Meteoritics 28:390–391

Masaitis VL, Pevzner LA (eds) (1999) Deep drilling in the Puchezh-Katunki impact structure. VSEGEI Press, St. Petersburg, p 392 (in Russian)

Masaitis VL, Danilin AN, Mashchak MS, Raikhlin AI, Selivanovskaya TV, Shadenkov EM (1980) Geology of astroblemes. Nedra Press, Leningrad, p 231 (in Russian)

Masaitis V, Naumov M, Ivanov B (1993) Fluidized ejecta of Puchezh- Katunki impact crater: possible implications to rampart craters of Mars (abs.). In: Abstracts for 16th International Russian-American microsymposium on planetology. Vernadsky Insititute, Moscow, pp 45–46

Masaitis VL, Mashchak MS, Naumov MV Selivanovskaya TV, Orlova JV (1995) The Puchezh-Katunki impact crater: main features of geological structure. Trans Dokl Russ Acad Sci 342:358–360 (in Russian)

Masaitis VL, Mashchak MS, Naumov MV (1996) The Puchezh-Katunki astrobleme: structural model of a giant impact crater. Sol Syst Res 30:5–13 (in Russian)

Mashchak MS, Masaitis VL, Kirjakov AF (1990) Main features of morphology and internal structure of the central uplift of the Puchezh-Katunki astrobleme (abs.). In: Abstracts for 21st all-union meteoritic conference. Moscow, pp 140–141 (in Russian)

Moldavskaya EP (1933) Geology and mineral resources of the former Nizhegorodsky Province. In: Mineral wealth of the Gorky region. Gorky Regional Press (in Russian)

Murchison RI (1845) The geology of Russia in Europe and the Ural mountains. J. Murray, London

Naumov MV (1993) Hydrothermal zonaton in central uplift of the Puchezh-Katunki astrobleme. Meteoritics 28:408–409

Naumov MV (2002) Impact-generated hydrothermal systems: Data from Popigai, Kara, and Puchezh-Katunki impact structures. In: Plado J, Pesonen LJ (eds) Meteorite impact structures in Precambrian shields. Impact studies, vol 2. Springer, Berlin, pp 117–171

Nechitailo SK, Veselovskaya MM, Skvortsova EN (1959) Materials for geology of the Gorodets-Kovernino tectonic zone. Gostoptekhizdat, Moscow, p 128 (in Russian)

Nikitin SN (1885) The summary geological map of European Russia, sheet 71 (Kostroma, Makarjev, Chukhloma). Proc Geol Comm 2(1) (in Russian)

Noinsky ME (1932) Upper Paleozoic deposits at the right bank of the Volga between Galashkino and Katunki settlements. VGRO Trans 51(6) (in Russian)

Osinski GR, Pierazzo E (eds) (2012) Impact cratering: processes and products. Wiley, Chichester, UK, p 336

Pevzner LA, Vorontsov AK, Kirjakov AF, Masaitis VL, Mashchak MS, Ivanov BA (1992) The Vorotilovskaya drillhole: the first deep drilling in the central uplift of large terrestrial impact crater (abs.). Lunar Planet Sci XXIII:1063–1064

Shatsky NS (1946) Coparative tectonics of ancient platforms, I: main features of structure and evolution of the East European Platform development. Proc Acad Sci USSR 1:5–62

Short NM, French BM (eds) (1968) Shock metamorphism of natural materials. Mono Book Corporation, Baltimore, MD

Shvetsov MS (1939) Geologic research of the southeast of the Sheet 72 of the geological map of the European Russia. In: Interior of the Gorky region, vol 2. GONTI Press (in Russian)

Sibirtsev NM (1896) Summary geological map of Russia in Europe, sheet 72 (Vladimir, Nizhny Novgorod, Murom). TransGeol Comm XV (in Russian)

Soloviev VK (1958) State geological map of the USSR on 1:1,000,000 scale. Sheet O-38 (Gorky). Gosgeoltekhizdat, Moscow (in Russian)

Susalnikova NV (1965a) On age of breccia and dislocated sequences from the Puchezh-Katunki dislocation area. Trans High Sch Geol Prospect 11:45–48 (in Russian)

Susalnikova NV (1965b) New data on the geological structure of the Gorodets-Kovernino tectonic zone. Soyuzburgaz Trans 5:55–69 (in Russian)

Tikhvinskaya EI (1956) Triassic glacial formations in the northern part of the Russian platform. In: Proceedings of the all-union conference for the designing of the united stratigraphic chart for Mezozoic of the Russian platform (in Russian)

Tumanov RR (1973) New data on the geological structure of the Gorodets-Kovernino tectonic zone. In: Nechitailo SK (ed) Materials on geology of the East Russian platform, vol 5, pp 112–125 (in Russian)

Tumanov RR, Bogorodskaya OA (1975) Geological map of the USSR on 1:200,000 scale. Sheet O-38-XXVI. VSEGEI, Leningrad (in Russian)

Ulanov EI, Pisannikova EL, Chumakov OE (2005) Legend of the state geological map of Russian Federation of 1:200,000 scale. Middle Volga series of sheets. Niznhy Novgorod, State Enterprise "Volgageologia" (in Russian)

Vardanyants LA (1961) A diatreme in the central part of the Russian platform. Trans Acad Sci Armen SSR 14(2):57–62 (in Russian)
Zemyatchensky PA (1886) Balakhna district. In: Records for estimate of lands of Nizhegorodskaya province. Nizhny Novgorod Provincial Zemstvo Press, Nizhny Novgorod (in Russian)

Geologic Setting of the Impact Site

Mikhail S. Mashchak and Mikhail V. Naumov

The Puchezh-Katunki impact crater formed in the central part of the East European platform. The impact site is two-layered: Archean to Paleoproterozoic crystalline basement is overlain by a platform cover consisting of carbonate and terrigenous Neoproterozoic, Paleozoic, and Lower Triassic rocks of ca. 2000 m thick. The most of the impact structure is buried beneath loose Mesozoic, Neogene, and Quaternary deposits up to 120 m thick.

Composition and structure of the basement are mainly known from the studies of cored material from the Vorotilovo Deep Well (VDW) and other deep boreholes (Fig. 1), from refraction and reflection seismic exploration surveys as well as from geological, gravimetric and magnetic surveys. The sedimentary cover has been investigated during medium-scale mapping, the detailed results of which are reported in unpublished Technical Reports of "Centergeologija" and "Volgageologija" by G. I. Blom, B. I. Fridman, R. R. Tumanov, A. V. Durandin, V. I. Belkevich, I. G. Kirkov as well as in published State Geological Maps of 1:200 000 and 1:1 000 000 scale (Kordun et al. 1975; Tumanov and Bogorodskaya 1975; Bogorodskaya and Tumanov 1980; Golts and Nazhestkina 1983; Fadeeva 2000). In addition, data from deep wells drilled outside the impact structure contributed to knowledge.

1 Crystalline Basement

The crystalline basement is opened by many deep boreholes both in Voritilovo Protrusion and outside of the impact structure. However, for the latter case, only the uppermost part of the basement was penetrated (for depth of no more than 50 m). The most of data on basement composition are obtained from numerous boreholes

M. S. Mashchak · M. V. Naumov (✉)
A.P. Karpinsky Russian Geological Research Institute,
199106 Sredniy Prospekt 74, St. Petersburg, Russia
e-mail: m_naumov@mail.ru; Mikhail_Naumov@vsegei.ru

V. L Masaitis and M. V Naumov (eds.), *The Puchezh-Katunki Impact Crater*,
Impact Studies, https://doi.org/10.1007/978-3-030-32043-0_2

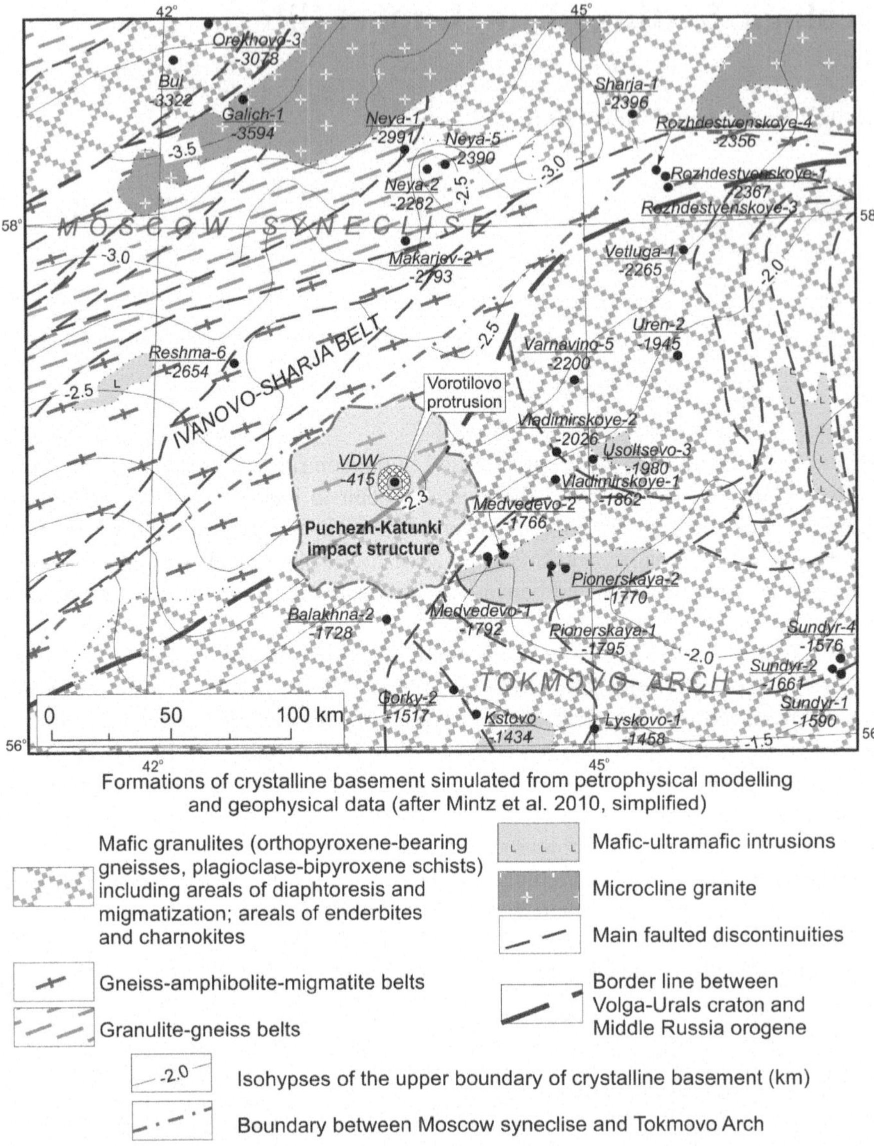

Fig. 1 Tectonic scheme of the crystalline basement of East European Platform in the area of Puchezh-Katunki impact structure. Formations of the basement simulated from petrophysical modeling and interpretation of geophysical data—after Mintz (2010). In addition, contour lines (in km) for the upper boundary of the basement are shown (compiled using unpublished technical reports by O. S. Bogatyrev, R. B. Davydov, K. Yu. Volkov and others). Deep wells penetrated into the basement are shown by black circles. The name of well is in numerator, the altitude of the upper boundary of the basement, in denominator. The central uplift of Puchezh-Katunki impact structure is hatched

in the central uplift of the Puchezh-Katunki impact structure, mostly from the VDW. The latter penetrated Miocene to Quaternary loose, alluvial and lacustrine deposits (from 0 to 62 m), clayey crater lake deposits (Kovernino Formation, from 62 to 265 m), coptomict gritstone and sandstone composed of re-deposited material of impact breccias and suevite (265–368 m), suevite (368–425 m), polymict impact allogenic breccia (425–550 m), and brecciated and shocked crystalline basement rocks from 550 to 5374 m (Masaitis and Pevzner 1999; Fig. 2). Thus, the VDW intersected 4824 m through crystalline basement lithologies. Although the latter are intensely transformed due to the impact event (brecciated, cataclased, shocked, annealed and altered by hydrothermal processes), original features of their composition and fabrics are mostly preserved.

Inferred from drilling and geophysical studies, the upper surface of the basement gradually dips in SSE–NNW direction from the most elevated part of the Tokmovo arch to the Moscow syneclise (Fig. 1). In the Balakhonikha well (130 km southeast of the VDW), the basement is opened at an attitude of −1181 m, while in the Galich-1 well (approximately 150 km northwest of the VDW), at −3595 m. When drilling and seismic profiling data from the outside of the impact structure having been extrapolated, a hypsometric position of the basement roof in the impact point is estimated to be about −2300 m. Thus, the basement inclines for 2500 m over the distance of about 300 km. The average incline within the Tokmovo arch is 3–5 m km^{-1}, but in the southeast slope of the Moscow syneclise it increases to 10 m km^{-1}. The basement rocks are folded, steeply dipping (60–80°) narrow isocline folds are of mostly nearly latitudinal strike. In many localities, sharp changes in the depth of the basement roof are revealed by seismic profiling and drilling to fix local features (lows and elevations) in the basement relief. Moreover, rupture dislocations of various range are present, which stipulated the blocky structure of the basement.

The impact site hit just over at the boundary zone between two mega-domains of the crystalline basement of the East European Platform, namely Volga-Urals gneiss-granulitic Archean continent and Middle Russian Paleoproterozoic mobile belt (Bogdanova et al. 2005). Inferred from geophysical data, these mega-domains differ radically in tectonic framework (Mintz 2010). The Volga-Urals craton is a complex compound of concentric features; such a feature is the Tokmovo ovoidal structure at the western flank of the craton. It is interpreted to be composed mainly of mafic granulites (orthopyroxene-garnet-plagioclase gneiss, plagioclase-bipyroxene schists) and enderbites. The tectonic framework of the Middle Russian orogene is dominated by ENE-WSW directed linear features. This mega-domain consists of alternation of granulite-gneiss and gneiss-amphibolite-migmatite belts aligned to its general strike. A southeastern frontier zone is considered as the Ivanovo-Sharja gneiss-amphibolite-migmatite belt, which is assumed to overthrust the Tokmovo structure (Mintz 2010). The belt is composed mostly of low-density lithologies, which may be assigned to biotite-amphibole gneiss with lenses and bands of amphibolites.

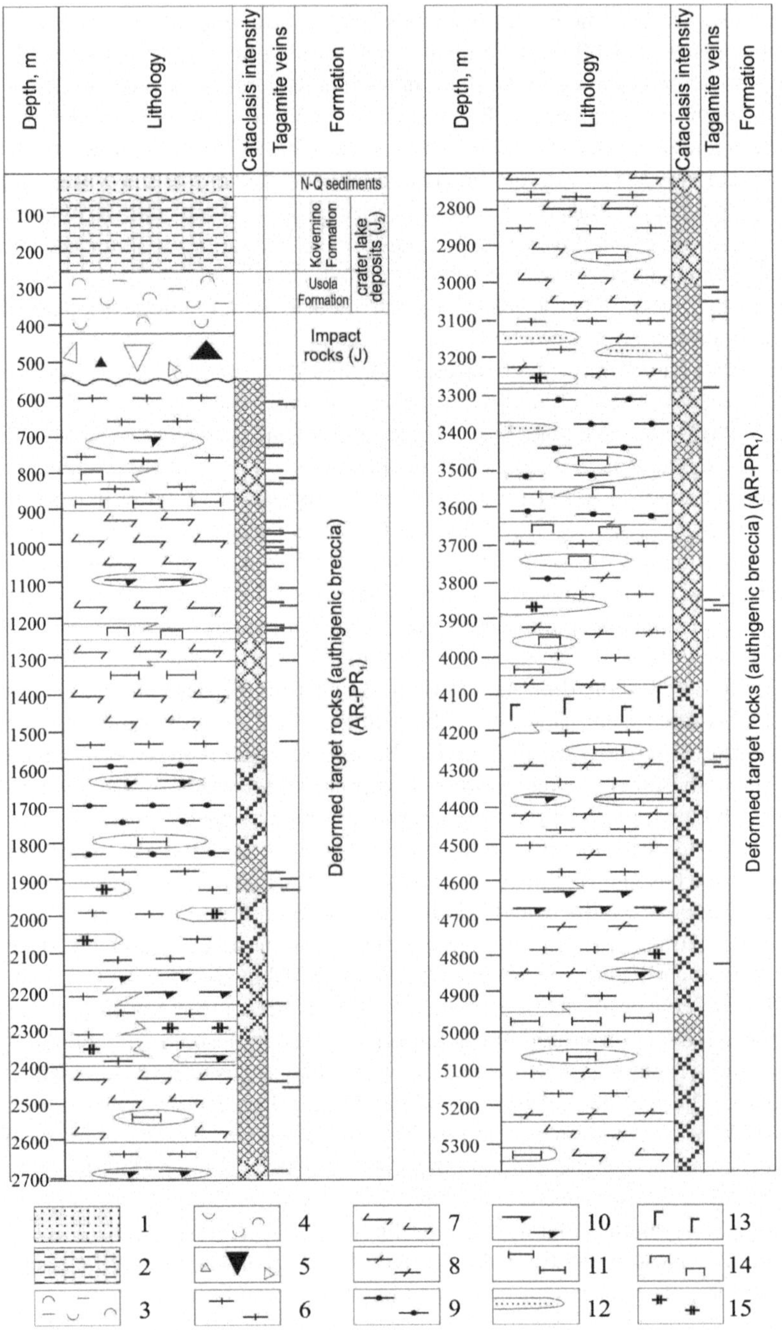

Depth, m
Lithology
Cataclasis intensity
Tagamite veins
Formation
N-Q sediments
Kovernino Formation
Usola Formation
crater lake deposits (J_2)
Impact rocks (J)
Deformed target rocks (authigenic breccia) ($AR-PR_1$)
100
200
300
400
500
600
700
800
900
1000
1100
1200
1300
1400
1500
1600
1700
1800
1900
2000
2100
2200
2300
2400
2500
2600
2700
2800
2900
3000
3100
3200
3300
3400
3500
3600
3700
3800
3900
4000
4100
4200
4300
4400
4500
4600
4700
4800
4900
5000
5100
5200
5300
1
2
3
4
5
6
7
8
9
10
11
12
13
14
15

◂**Fig. 2** Geological section of the Vorotilovo Deep Well. Lense-like bodies are shown out of scale (Masaitis and Pevzner 1999). 1—Sand, 2—clay, 3—coptomict gritstone, 4—suevite, 5—polymict allogenic breccia, 6—amphibole-biotite gneiss, 7—biotite-amphibole and amphibole gneiss, 8—pyroxene-amphibole and pyroxene-biotite-amphibole gneiss, 9—garnet-biotite, garnet-amphibole-biotite gneiss, 10—amphibolite, 11—plagioclase-bipyroxene schist, 12—quartzite including quartzitic gneiss, 13—dolerite, 14—ultramafics, 15—granite and granite gneiss. The intensity of cataclasis and brecciation of rocks is shown by hatching density (the denser hatching corresponds to more intense deformation). In addition, depth intervals of tagamite (impact melt rock) veins distribution are conditionally shown

1.1 Metamorphic Lithologies

Crystalline basement lithologies opened by boreholes in the vicinity of the Puchezh-Katunki impact structure (within a radius of 100 km away from the Vorotilovo protrusion) are mainly biotite-amphibole gneisses and migmatites (wells Reshma-6, Makarjev-2, Balakhna-2, Gorky-2 etc.). However, the uppermost part of the basement east of the impact structure is represented by biotite-pyroxene gneisses, plagioclase-bipyroxene schists, or enderbites (wells Varnavino-5, Uren-2, Pionerskaya-1, 2, Medvedevo-1, 2 located at 20–50 km from the edge of the impact structure; *see* Fig. 1). This difference reflects a complicated tectonic position of the area, which is situated at the turn of two distinct mega-domains. However, a great bulk of information on metamorphic lithologies at the impact site was provided by numerous cores from the Vorotilovo protrusion where 62 boreholes penetrated into the basement.

Although the most of these boreholes uncovered an uniform sequence composed of leucocratic to mesocratic amphibole-biotite-plagioclase gneisses with amphibolite lenses, the VDW section showed that the basement at the impact site comprises a wide variety of metamorphic lithologies. The latter may be grouped into eight associations based on their mineral and chemical composition and fabrics: 1—biotite, amphibole-biotite, biotite-amphibole and amphibole gneisses; 2—pyroxene-amphibole-biotite, biotite-pyroxene-amphibole and amphibole-pyroxene gneisses with garnet, titanite, and epidote; 3—high-alumina (biotite-garnet, biotite-amphibole-garnet, biotite-kyanite-sillimanite-garnet) gneisses; 4—quartzites and quartzitic gneisses; 5—calciphyre, marbles, skarns; 6—amphibolites; 7—amphibole-pyroxene schists; 8—phlogopite-anthophyllite schists. All lithologies are described in detail in Masaitis and Pevzner (1999). Chemical compositions of different basement lithologies are presented in the AFM diagram (Fig. 3); their average chemical compositions are given in Table 1.

Biotite-amphibole, amphibole-biotite, and amphibole gneisses vary essentially by mineral composition. The rocks consist of intermediate oligoclase (An_{26} on average, 35–60%), quartz (10–35%), biotite (0–25%), microcline (0–30%), and ferriferous hornblende (0–25%). Apatite, zircon, ilmenite, magnetite do occur but they contribute as low as tenths of a percent. All varieties are clearly gneissic with variable percentage of amphibole, biotite, and plagioclase. Biotite-amphibole and, particularly, amphibole gneisses commonly are more melanocratic, coarse-grained and,

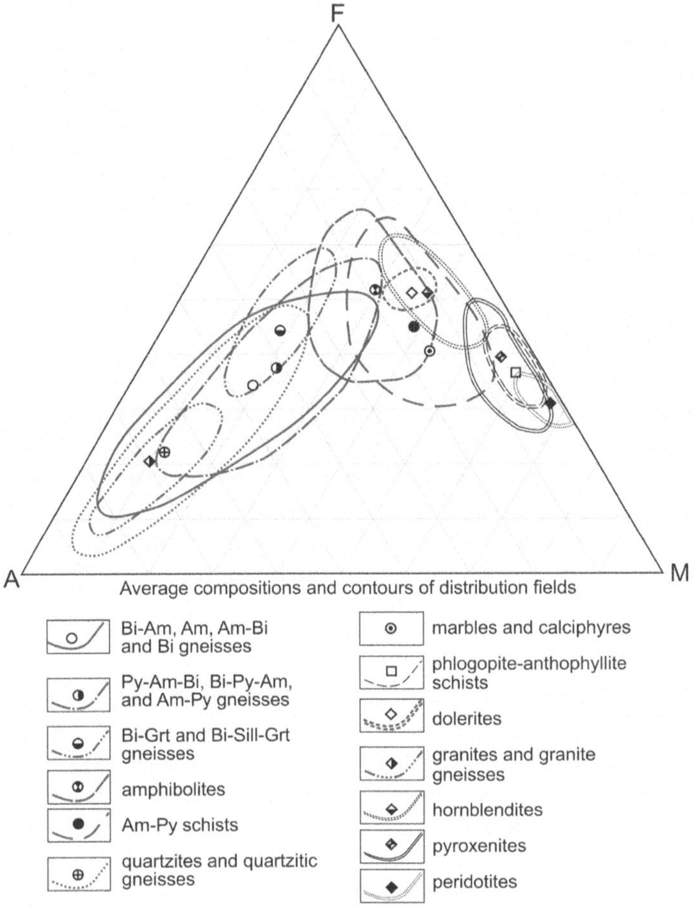

Fig. 3 Ternary AFM diagram for crystalline basement lithologies from the VDW section. The average weighted compositions of the lithologies are given in Table 1

frequently, massive. Biotite gneisses are usually enriched by microcline and quartz due to migmatization. Unlike mineral assemblages, the chemical composition of considered rocks is persistent. Noticeable variations have been recorded only for silica, iron, and magnesium contents (Table 1).

Pyroxene-amphibole-biotite, biotite-pyroxene-amphibole, amphibole-pyroxene gneisses differ from the above group by a permanent clinopyroxene occurrence and predominant mesocratic composition. Contents of rock-forming minerals vary appreciably. Plagioclase is acidic andesine (An_{33} on average), it contributing 35–60%, quartz, 3–20%, hornblende, 5–40%, magnesium salite ($En_{40}Fs_{14}$ on average), 1.5–25%, biotite, 0.5–10%. Garnet ($Alm_{61}Pyr_{12}Gros_{24}Spes_{3}$), titanite, carbonate, and clinozoisite contribute variably, but usually no more than 5–6%. Magnetite, ilmenite, muscovite, chlorite, apatite, zircon are accessory minerals. Microcline can

Table 1 Average chemical composition (wt%) of crystalline basement lithologies

Lithology	1		2		3		4				5				6		7	
							a		b		a	b	c	d				
	X	s.d.	X	s.d.	X	s.d.	X	s.d.	X	s.d.					X	s.d.	X	s.d.
SiO_2	63.52	3.83	61.65	4.33	59.57	4.70	87.89	3.98	71.40	1.62	45.73	36.51	30.18	43.94	49.91	2.98	48.74	3.43
TiO_2	0.58	0.22	0.68	0.25	0.82	0.30	0.08	0.07	0.18	0.08	0.76	1.59	0.14	1.33	1.36	0.52	1.37	0.93
Al_2O_3	15.12	0.99	15.14	1.29	17.90	1.90	3.56	2.26	13.23	1.72	12.06	6.88	1.19	12.48	13.83	1.25	11.96	2.12
Fe_2O_3	2.03	0.98	1.64	0.85	1.52	1.22	0.76	0.45	0.78	0.54	1.75	3.43	4.50	2.92	4.67	1.83	3.09	1.29
FeO	2.63	1.29	3.89	1.95	5.36	1.45	1.89	1.59	1.08	0.44	4.62	7.88	6.61	9.98	7.29	2.60	8.32	2.25
MnO	0.09	0.13	0.15	0.30	0.11	0.03	0.04	0.03	0.04	0.02	0.10	0.10	0.21	0.21	0.19	0.06	0.20	0.05
MgO	2.45	1.19	3.04	1.38	2.81	1.02	0.84	0.42	1.08	0.90	8.57	12.31	32.57	8.40	6.83	2.01	9.27	3.51
CaO	4.15	1.31	4.97	2.04	3.81	1.88	1.80	1.15	3.02	1.83	20.13	15.40	10.05	16.18	8.91	1.44	10.80	2.26
Na_2O	4.40	0.74	4.19	1.09	3.71	1.56	0.18	0.19	3.66	1.12	0.73	0.00	0.13	1.07	2.98	0.67	2.43	0.90
K_2O	2.05	0.89	1.84	0.77	2.23	0.82	0.23	0.10	3.50	1.31	1.23	1.04	0.68	1.45	1.07	0.46	1.15	0.56
P_2O_5	0.19	0.12	0.18	0.15	0.22	0.29	0.02	0.03	0.06	0.05	0.23	3.31	0.00	0.00	0.19	0.14	0.31	0.35
L.O.I.	2.72	1.37	2.25	0.85	1.72	0.70	1.99	0.92	1.82	0.44	3.87	11.40	13.62	1.77	2.60	1.61	2.03	1.00
Totals	99.93		99.61		99.79		99.27		99.84		99.78	99.85	99.88	99.73	99.82		99.66	
n	130		42		12		4		17		1	1	1	1	81		38	

Lithology	8		9		10						11		12				13	
					a		b		c				a		b			
	X	s.d.	X	s.d.	X	s.d.	X	s.d.	X	s.d.	X	s.d.	X	s.d.	X	s.d.	X	s.d.
SiO_2	48.44	2.83	39.14	4.07	45.71	1.86	46.15	2.29	44.28	1.49	45.59	2.33	68.54	2.62	71.77	2.04	49.36	0.87
TiO_2	0.64	0.03	0.71	0.21	0.97	0.58	1.13	0.49	1.88	0.66	2.76	1.12	0.37	0.12	0.17	0.09	1.21	0.07

(continued)

Table 1 (continued)

Lithology	8		9		10						11		12				13	
					a		b		c				a		b			
	X	s.d.	X	s.d.	X	s.d.	X	s.d.	X	s.d.	X	s.d.	X	s.d.	X	s.d.	X	s.d.
Al_2O_3	7.55	0.78	2.02	0.97	7.91	1.29	8.48	0.81	7.68	1.18	10.19	1.86	15.06	1.22	13.70	1.10	14.46	0.38
Fe_2O_3	4.60	1.27	7.92	1.71	4.80	0.90	6.20	3.00	4.93	1.75	4.25	1.05	0.82	0.26	0.68	0.39	3.07	0.53
FeO	6.81	0.73	6.81	1.66	7.75	1.59	5.76	2.58	8.45	1.34	10.53	2.62	1.59	0.47	0.94	0.31	7.94	0.62
MnO	0.13	0.02	0.21	0.06	0.19	0.04	0.18	0.04	0.18	0.02	0.23	0.07	0.03	0.03	0.02	0.02	0.17	0.02
MgO	18.92	0.69	32.91	4.26	20.36	1.54	18.00	3.03	16.70	1.29	10.53	2.59	1.19	0.34	0.82	0.35	7.44	0.51
CaO	3.65	2.39	1.74	0.76	7.72	2.37	8.24	1.34	11.35	1.61	10.94	1.70	3.11	0.84	2.02	0.57	10.66	1.15
Na_2O	0.44	0.00	–	–	0.82	0.49	1.26	0.57	1.11	0.24	1.87	0.53	4.94	0.49	3.76	0.88	1.89	0.28
K_2O	0.78	0.04	0.58	0.65	0.86	0.62	0.77	0.48	0.45	0.31	0.92	0.53	2.23	0.47	4.13	0.80	0.77	0.34
P_2O_5	0.06	0.00	0.09	0.03	0.09	0.10	0.19	0.13	0.23	0.19	0.30	0.17	0.11	0.10	0.03	0.04	0.11	0.02
L.O.I.	8.20	0.34	7.94	2.92	2.49	0.83	3.62	2.20	2.38	0.80	1.68	0.94	1.74	0.84	1.84	0.56	2.71	1.25
Totals	100.21		100.06		99.67		99.97		99.64		99.80		99.73		99.88		99.79	
n	3		6		9		8		11		9		13		11		15	

XRF data, analyzed by B. A. Tsimoshenko, VSEGEI Labs, St Petersburg, with a SRM-2 instrument

X Mean value; *s.d.* Standard deviation; *n* Number of samples analyzed

1—Amphibole-biotite, biotite-amphibole, and amphibole gneisses; 2—pyroxene-bearing gneisses: 3—high-alumina gneisses; 4a—quartzites, 4b—quartzitic gneisses; 5—calciphyre: 5a, phlogopite-scapolite-forsterite, 5b, apatite-diopside, 5c, forsterite-diopside, 5d—skarn; 6—amphibolites; 7—plagioclase-bipyroxene schists; 8—anthophyllite-phlogopite schists; 9—peridotites; 10a—olivine websterites, 10b—websterites, 10c—diallagite; 11—hornblendites; 12a—biotite tonalites, 12b—two-mica microcline granites and granite gneisses; 13—dolerites occupy up to 20% of total rock volume. Clinopyroxene occurs mainly as relics against hornblende; the latter is more ferriferous than in biotite-amphibole gneisses. Biotite is partially or completely replaced by pennine and, more rarely, muscovite. Chemical composition of the group is rather persistent irrespective of variable mineral composition (Table 1).

High-alumina (biotite-garnet, biotite-amphibole-garnet, biotite-sillimanite-kyanite-garnet) gneisses are gneissic and banded rocks with porphyroblastic texture. The mineral composition is dominated (30–60%) by plagioclase, which varies from andesine to basic oligoclase (An_{25} on average), quartz contributes 2–10%, garnet of variable composition ($Alm_{54-67}Pyr_{8-22}Gros_{8-32}Spes_{3-6}$), 5–25%, biotite, 3–15%, hornblende, 0–15%, kyanite and sillimanite, up to 10–15%. graphite and salite, up to 2–5%. Titanite and magnetite are accessory. Microcline contributes 10–15% in granitized and migmatized varieties. Chemical compositions of these lithologies are distinguished by high contents of alumina and iron (Table 1).

Quartzites are nearly monomineral rocks, 1–2% of clinopyroxene, biotite, epidote, muscovite, chlorite, magnetite, or sillimanite can occur. *Quartzitic gneisses* originate due to migmatization of quartzites; they form separated bands in the latter. This lithology is distinguished by the gneissic structure and high content of biotite and feldspars; the latter may contribute as much as 35%.

Marbles, calciphyres and skarns are calc-silicate rocks formed lenses within pyroxene-bearing gneisses. Pure marbles and calciphyres have not been recorded. Commonly, an alternation of bands and spots of marble, calciphyres, skarns, basic schists, or amphibolites occur. The mineral composition is complicated in places by migmatization and granitization.

Marbles consist of calcite or dolomite (from 70 to 95%), feldspars (from 3 to 15%), clinopyroxene, hornblende, muscovite, talc, titanite, and graphite (from 1–2 to 10–15% together), hematite and magnetite being minor. Mineral composition of calciphyres is much more diverse and variable: along with ubiquitous calcite and diopside contributing from 5 to 60 vol.%, various proportions of feldspars (predominantly plagioclase), scapolite, garnet, hornblende, phlogopite, talc, muscovite, serpentine, apatite, forsterite (or products of its alteration), spinel, titanite, graphite occur. So, phlogopite-scapolite-diopside, apatite-diopside, forsterite-diopside varieties of calciphyre and garnet-titanite-scapolite-diopside scarns may be distinguished.

Amphibolites are massive or banded medium- to coarse-grained rocks, which form lenses and bands within various gneisses; the maximum thickness of an amphibolite is 65 m in core. The lithology is dominated by hornblende (50–80%) and andesine (An_{34} on average) (5–40%). Subordinate minerals occur in many bodies, they are clinopyroxene ($En_{35}Fs_{20}$, up to 10%), quartz, up to 5%, biotite, up to 3%, garnet, up to 5%. Titanite, apatite, epidote, magnetite, titanomagnetite, and ilmenite are minor constituents; however, iron and titanium oxides peak 5% in some amphibolites. Hornblende is characterized by a high alumina content; it is replaced frequently by chlorite and biotite; in places, it hosts salite relics. Similar to gneisses, amphibolite underwent migmatization in many places; this process is reflected by substitution of plagioclase by potassium feldspars and increasing of quartz.

Amphibole-pyroxene schists form lenses peaking a few meters thick within pyroxene-bearing gneisses or amphibolites. This is massive lithology consisting of magnesium salite ($En_{40}Fs_{15,}$ 30–50%), hornblende (25–40%), and plagioclase (up to 15–20%). The latter ranges in composition from oligoclase to acidic andesine indicating its origin due to migmatization. Quartz may contribute up to 3–5% in migmatized parts; in places, almandine garnet occurs peaking 10%, while alteration carbonate may reach 5%. Titanite, apatite, ilmenite, and biotite are minor. Clinopyroxene is replaced frequently by hornblende and chlorite. Hornblende is high-magnesium and low-alumina like pyroxene.

Phlogopite-anthophyllite schists are metamorphosed ultramafics. In addition to predominant anthophyllite and subordinate phlogopite, high-chromous and nickeliferous magnetite and high-magnesium ilmenite contribute up to 3%. Relict olivine and orthopyroxene occur in places.

The described metamorphic lithologies distribute unevenly in the VDW section, so that two series differing in composition and metamorphic overprint may be distinguished (Masaitis and Pevzner 1999). The Upper series is opened in the depth interval of 550–3080 m. It is represented by an uniform alternation of leucocratic to mesocratic amphibole-biotite, amphibole, and biotite gneisses, which contribute more than 65 vol.% The rest of the series is composed of amphibolites (17%), biotite-amphibole-pyroxene gneisses (10%), plagioclase-bipyroxene schists (5–7%), high-alumina gneisses (mainly of biotite-sillimanite-garnet composition at the depth interval of 1500–1800 m; 3%). All lithologies underwent to migmatization and granitization to a variable degree. Granite, granite gneisses and quartzites contribute 3–5%. Within some depth intervals (770–840 and 1150–1350 m), thin ultramafic bodies occur within gneisses and amphibolites, they contributing no more than 0.3% to the sequence. All other boreholes in the Vorotilovo protrusion penetrated the same sequence, pyroxene-bearing gneisses were found only within the most continuous uncovered sections (706 m, bhs. 13 and S-3).

The Lower series (depths of 3080–5374 m) is more heterogeneous, It is composed mainly of pyroxene-biotite-amphibole and amphibole-pyroxene gneisses contributing 55 vol.%; these lithologies contain frequently almandine garnet, titanite, clinozoisite. Within pyroxene-bearing mesocratic gneisses, lenses of amphibole-pyroxene schists appear below 4000 m. Amphibolites (20%) distribute unevenly, they occur preferably below 3500 m. Biotite-amphibole gneisses (7%) are subordinate. In addition, high-alumina gneisses (of biotite-garnet and biotite-sillimanite-kyanite-garnet composition) contribute appreciably (6%), but they are confined to the depth interval of 3200–4000 m. Acidic rocks—granites and granite gneisses, quartzites and quartzitic gneisses contribute more than 10%. In addition, calc-silicate lithologies (calciphyres, marbles, and diopside-scapolite scarns) appear below 4350 m, they contributing up to 2% for this depth interval. Igneous lithologies—dolerites and ultramafics contribute less than 1%.

The weighted average compositions of Upper and Lower series are given in Table 2. The difference between two series does not turn so significant as if it should be followed from mineral composition of main lithologies. This may be explained by the common granitization of both series. Nevertheless, the average composition

Table 2 Average weighted chemical compositions (wt%) of Upper and Lower series of the metamorphic sequence drilled by the VDW

Components	Upper series	Lower series
SiO_2	60.33	59.05
TiO_2	0.73	0.71
Al_2O_3	14.50	13.70
Fe_2O_3	2.56	2.28
FeO	3.79	4.51
MnO	0.11	0.13
MgO	4.13	5.73
CaO	4.94	5.19
Na_2O	3.88	3.37
K_2O	1.85	1.76
P_2O_5	0.18	0.21
L.O.I.	2.90	3.31
Totals	99.90	99.97
n	351	206

n Number of samples analyzed

of the Lower series differs from that of the Upper one by higher magnesium and iron values and low silica, alumina, and alkali values.

Inferred from metamorphic rock associations, the Upper series is a gneiss-amphibolite-migmatite complex, which is typical for the southernmost part of the Middle Russia orogene (Ivanovo-Sharja belt). The Lower series could be considered rather as a diaphtorized granulite-enderbite-gneiss association. The latter is not only widespread in the northwestern flank of the Volga-Urals craton southwest of the Puchezh-Katunki, but it also forms intensely transformed inliers within gneiss belts of the Middle Russian orogene (Fig. 1).

Throughout the Vorotilovo protrusion including the VDW section down to 5374 m, crystalline basement rocks are brecciated and cataclased. Therefore, cataclasites, mylonites, and pseudotachylites are widespread, especially at the depth above 3000 m; downward in the section, cataclastic lithologies occur only within relatively thin (up to 40 m in core) zones. This indicates that a major portion of cataclasites originated due to impact-generated transformation of the crystalline basement. However, some of them can be older. There are no clear criteria to distinguish cataclastic formations arisen during Early Precambrian dynamothermal metamorphism from those that were derived from the impact. If these rocks retaining signs of shock transformations, their origin is evident, though, the most of cataclasites are lacking undoubted shock features.

The age of the metamorphic sequence is accepted to be Neoarchean based on U-Pb dating of zircons from various gneisses occurring at different deep levels of the VDW section (Naumov et al. 2015). Two main events are revealed: (1) 2540–2530 Ma—formation of the gneiss-amphibolite Neoarchean sequence under *PT* conditions of the amphibolite facies; (2) 2020–1980 Ma—anatexis and granitization of the Neoarchean amphibolite-gneiss complex that appeared as a series of discrete bursts of activity. The Neoarchean age of gneisses is confirmed by the neodymium model age of a gneiss sample calculated relative to the depleted mantle (T^{Nd}_{DM}). This value is 2.65 Ga, constraining the time of the protolith formation (Masaitis and Pevzner 1999).

1.2 Igneous Lithologies

Igneous lithologies opened by the VDW include ultramafics, granitic rocks, and dolerites.

Ultramafics form sub-conformable, relatively thin sheet bodies within the amphibolite-gneiss sequence. Among them, pyroxenites and hornblendite have been recorded at different levels of the metamorphic sequence (770–840, 1150–1350, and 3550–4100 m).

Pyroxenites peak 23 m thick in core. They are composed of augite or magnesium salite (45–80%), orthopyroxene (En_{67-83}, 0–30%), forsterite (Fo_{78} on average, 0–15%), and magnetite (2–5%). All pyroxenites are amphibolized to a variable degree; magnesium pargasite (15–50%) and phlogopite (0–20%) contribute essentially. Other alteration minerals such as anthophyllite, tremolite, talc, serpentine, and biotite occur in addition. Three varieties—websterite, olivine websterite, and diallagite may be distinguished based on mineral composition. Chemical composition of pyroxenites is characterized by lower magnesium and higher alumina and iron contents (Table 1) as compared with standard composition for this lithology that is most likely due to dynamothermal metamorphism.

Hornblendites associate with pyroxenites and include frequently lenses and boudins of the latter, rare small separate hornblendite bodies are also encountered, though. Hornblendite lenses do not exceed 8 m thick. The main rock-forming mineral, hornblende (75–90%), is represented by two distinct varieties: magnesian pargasite (magnesium number is about 78) similar to that developing after clinopyroxene in pyroxenites, and ferriferous hornblende (magnesium number is about 47), which is characteristic of metamorphic amphibolites. Relict augite (up to 5%) is common as well as ilmenite and magnetite. In metamorphosed, gneissic varieties, biotite and phlogopite may reach 10 to 15%, some feldspars appear in addition. Anthophyllite, tremolite, talc, and serpentine are found in places.

As contrasted to pyroxenites and hornblendites, peridotites occur only within a short depth interval in the upper metamorphic series (between 790 and 820 m). They form thin (up to 0.6 m thick) boudins. Original minerals (olivine, clinopyroxene, and orthopyroxene) remain only as relics. The rocks consist mostly from serpentine (70–90%), which is scattered by ore minerals (high-magnesium and high-chromous magnetite and high-magnesium ilmenite) grains. Talc, phlogopite, spinel, and sulfides occur in addition. In some bodies, magnesium hornblende contributes up to 30%. All original minerals are high-magnesium (magnesium number is 67 for augite, 79 for orthopyroxene, and 80 for olivine). The most of peridotites are transformed to phlogopite-anthophyllite schists, which are present within a slightly longer depth interval in VDW section (770–840 m).

The age of all ultramafics is accepted to be Neoarchean because of these lithologies are intensely diaphtorized and partly foliated like host metamorphic rocks.

Granites form some bodies up to 30 m thick that are intersected by the VDW within depth intervals of 1.8–2.5 and 3.2–4.5 km. Biotite tonalite and microcline two-mica granite are distinguished. The former consists of oligoclase (An_{23} on average) (40–60%), microcline (10–35%), quartz (15–25%), and biotite (1–3%). Zircon, magnetite, and ilmenite are accessory minerals..

Microcline two-mica granites together with migmatites constitute the major portion of granitoids. These are gneissic medium-grained rocks dominated by alkali feldspars: altered orthoclase and fresh of microcline-microperthite (55–75%). Quartz contributes 15–25%, sericitized oligoclase, up to 15%, biotite, muscovite, and clinozoisite, no more than 1% in totals. In addition, minor zircon, apatite, and magnetite occur. The chemical composition of both granite lithologies is wide-ranging.

Based on mode of occurrence, relationships with host rocks, variable composition, all granites from the VDW section are interpreted to origin by anatexis of host metamorphic rocks. The age of a microcline granite measured by U-Pb dating of zircons is about 1985 Ma (Naumov et al. 2015).

Dolerite. Several dolerite dikes are opened by the VDW at depth interval of 3780–4320 m; they range from 0.2 to 20 m in core. Dolerites are uniform, they consist of labradorite (45–65%), augite (15–30%), alteration chlorite (5–15%), titanmagnetite (2–5%), and tholeiitic mesostasis (up to 10%); minor titanite and ilmenite occur in addition. The U-Pb dating of zircon grains from dolerites by SHRIMP-II techniques showed their Neoarchean age (about 2530–2540 Ma); however, these zircons are interpreted to be foreign (Naumov et al. 2015). Analogous dolerites cutting the basement in the Volga-Urals craton are dated by K-Ar techniques from 1150 to 1100 Ma (Bogdanova and Lapinskaya 1989; Muslimov and Lapinskaya 1996). The Mesoproterozoic age of dolerites from the VDW section seems to be more probable.

2 Sedimentary Cover

The impact structure is situated at the transition zone between two major tectonic features of the Russian plate, Volga-Urals anteclise (Tokmovo Arch) at southeast and Moscow syneclise at northwest. A boundary location between these tectonic features is disputed, but it is commonly outlined in SW-NE direction close to the northwest edge of the impact structure where a benched increasing of the thickness of Neoproterozoic deposits occurs (Fadeeva 2000). The platform cover in the impact site comprises Neoproterozoic, Devonian, Carboniferous, Permian, and Lower Triassic sedimentary rocks of ca. 2000 m total thick. From the northwestern slope of the Tokmovo arch (southeast of the impact structure) toward the central part of the Moscow syneclise the thickness of the platform cover increases from 1400 to more than 3000 m (Fig. 1). A gradual change in the thickness of the platform cover and its cutoff by the impact may be demonstrated by a S–N directed profile through deep boreholes Gorky–Balakhna–VDW–Makarjev (Fig. 4). At the impact structure

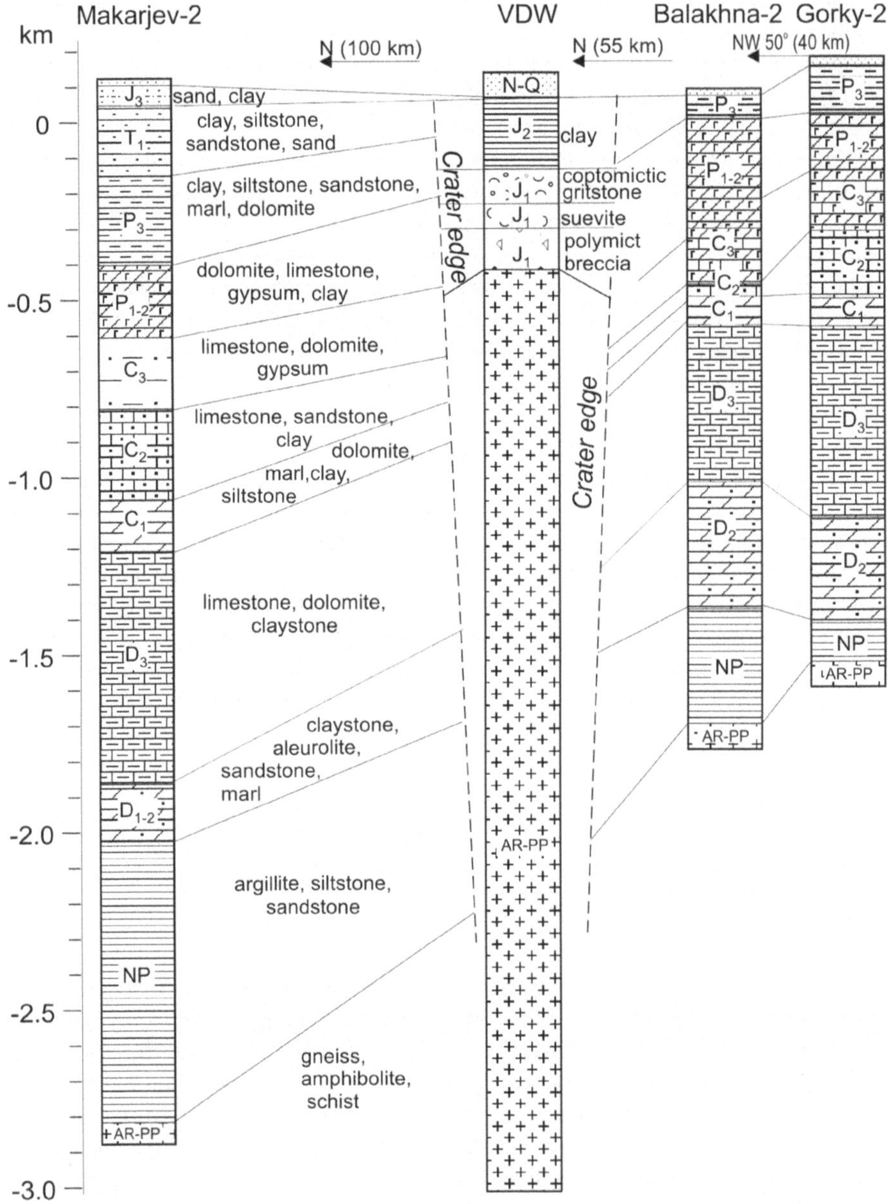

Fig. 4 Correlation scheme for some well sections in areas adjacent to the Puchezh-Katunki impact structure. These wells demonstrate a N–S directed profile. Location of the wells is given in Fig. 1

proper, this thickness ranges from 1600 to 2300 m.

The stratigraphy of the region is studied in detail due to the extensive deep drilling and based on complex fossil investigations, results of which are synthesized in many summary papers (Tumanov and Bogorodskaya 1975; Fadeeva 2000; Ulanov et al. 2005; Kirikov 2014; Kirkov et al. 2019 and many others). Thus, the platform cover column in the impact site may be recovered with confidence (Fig. 5). The cover includes two stratigraphic successions: Neoproterozoic and Middle Paleozoic—Lower Mesozoic that are separated by a continuous gap in the deposition.

Neoproterozoic deposits are designed to range from 350 to 700 m (ca. 450 m in the place of the present crater center). Its upper boundary rises northwesterly from -1.7 to -1.4 km. The Neoproterozoic is characterized by platform-type smooth dislocations of mostly latitudinal strike; the beds dip at angles of 1–4° in places.

The Neoproterozoic rocks are opened by numerous boreholes in the Vorotilovo Protrusion where they are brecciated and dislocated. The nearest undisturbed Neoproterozoic sections are penetrated by wells Balakhna-2 (328 m thick) at 50 km south of the center of the Puchezh-Katunki crater and Reshma-6 (764 m thick) at 85 km northwestward. The Neoproterozoic sequence at the impact site is incomplete compared to the typical section for Moscow syneclise; it is represented only by the Gavrilov Jam Formation (middle part of the Redkino Group; Fadeeva 2000). It consists mainly of bluish-grey or greenish-grey argillite with siltstone interlayers up to 11 m thick and rare, thin (<0.5 m) sandstone, limestone, and shale interbeds. Close to the base of the sequence, red and brown argillite are predominant.

The Middle Paleozoic–Lower Mesozoic succession involves a thick sequence of both shallow marine and continental deposits of Devonian, Carboniferous, Permian, and Lower Triassic. It rests with angular and stratigraphic unconformity on the Neoproterozoic and varies in thickness from 1350 (at the southeastern edge of the crater) to 1700 m (at its northwestern edge). Frequent stratigraphic gaps and erosion surfaces have been recorded inside the succession. The bedding planes commonly are close to horizontal position. Amplitudes of local positive and negative structures (elongated structural heights, depressions, local upheavals, flexures etc.) do not exceed 30 m. In contrast to the predominantly latitudinal and northwestern directions of tectonic elements characteristic of the Neoproterozoic succession, the Middle Paleozoic–Lower Mesozoic one is characterized by meridional structural features.

The succession considered is dominated by Devonian carbonate and terrigenous sediments of 820–850 m thick. The Devonian sequence includes deposits of all three series and looks as follows.

Lower Devonian. Emsian stage (ca. 20 m thick): light-grey quartz sandstone with siltstone interlayers.

Middle Devonian. Eifelian stage (ca. 85 m thick): Grey dolomite and limestone with claystone and marl interlayers. Givetian stage (ca. 107 m thick): Three rhythmites consisting of light-colored quartz sandstone in lower parts, with transition to siltstone close to the top.

Upper Devonian. Fransian Stage (ca. 400 m thick) is composed in the lower part of siltstone with interlayers of sandstone and, close to top, limestone (up to 100 m thick); the most of this stage (up to 320 m) consists of greenish-grey to grey limestone

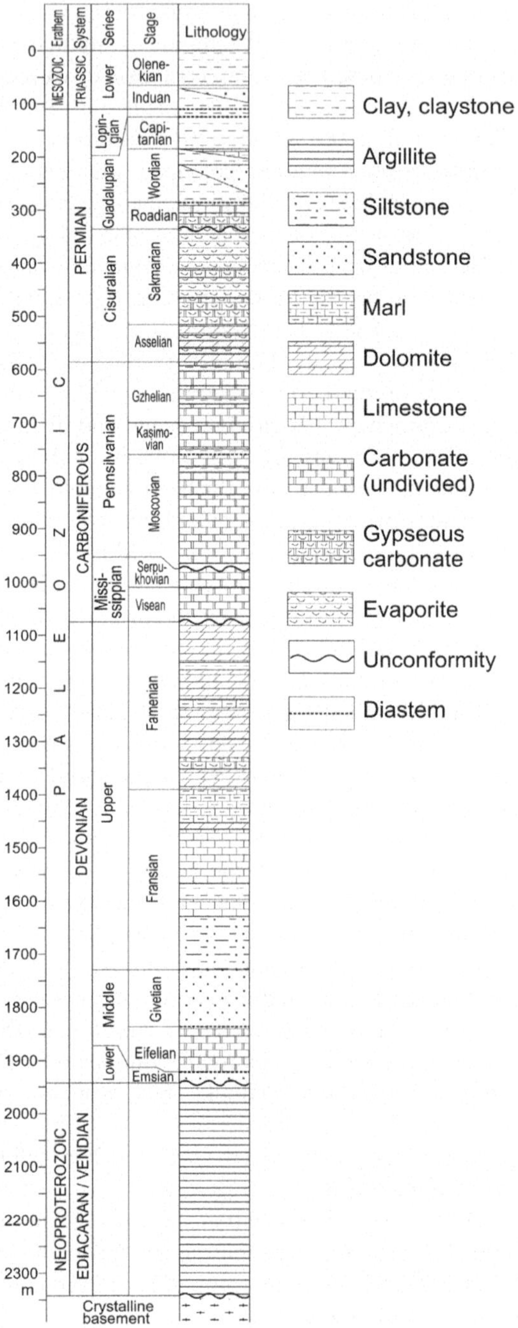

Fig. 5 A restored stratigraphic column of the platform cover at the impact site. The column is based on drilling data on distribution of different stratigraphic units in areas adjacent to the Puchezh-Katunki impact structure (Tumanov and Bogorodskaya 1975; Fadeeva 2000; Kirkov et al. 2019, and others)

with marl and claystone interlayers, the latter peaking 32 m thick. Famenian stage (up to 224 m thick): yellowish-grey to grey dolomite with marl and claystone interlayers and lenses of gypsum and anhydrite.

The Carboniferous system is represented by shallow marine, coastal, lagoon, and continental sediments attributed to all three series. The summary thickness of the Carboniferous is estimated to be 450 m. Between Devonian and Carboniferous, a stratigraphic gap occurs including Tournaisian and, partly, Visean stages. The structural framework of the Carboniferous differs considerably from that of Devonian.

Mississippian. Upper Visean Stage (ca. 55 m). At the base, a thin (9 m) horizon of black shale with clayey dolomite interlayers lies over erosion surface of the Devonian. The bulk of the stage consists of yellowish-grey and white limestone and dolomite with rare claystone interlayers. Lower Serpukhovian Stage (35 m) is represented by white limestone and dolomite.

Pennsylvanian is represented by Moscovian, Kasimovian, and Gzhelian Stages. The Moscovian overlies Mississippian deposits with ravinement. Its sequence comprises, from bottom to top: variegated clays with dolomite and siltstone interlayers (33 m); dolomites intercalated with limestones (from 120 to 145 m); limestones with subordinate dolomite and chert, clay, and marl lenses (58–90 m). The thickness of the Moscovian stage increases northerly.

The Kasimovian Stage (from 45 to 80 m thick) is dominated by light-grey limestones with subordinate dolomites and rare chert and gypsum lenses; a thin (3–5 m) variegated clay horizon occurs in the base. The uppermost of the Carboniferous (Gzhelian Stage) represents a rhythmic unit consisting of interlayered dolomite and limestone, with regular clay interlayers. Its summary thickness varies highly (from 87 to 155 m).

Permian deposits overlie conformably the Pennsylvanian. Its thickness varies from 185 m close to the southwest edge of the Puchezh-Katunki to 450 m outside of its northeastern edge. Thus, in contrast to the Devonian, the lower boundary of the Permian submerges northeasterly. South and southwest of the impact structure, Permian deposits are exposed in stream valleys from under impact breccias.

The Cisuralian continues the Pennsilvanian shallow marine sequence but differs by a significant occurrence of evaporites. It is represented only by its lower part (Asselian and Sakmarian Stages). The sequence consists mainly of dolomite with anhydrite and gypsum interbeds; in places, limestones intercalate with dolomites. The upper part of the sequence (Sterlitamak horizon) peaking as much as 141 m thick is composed of bluish anhydrite with rare dolomite lenses and nodules; in addition, a thin (0.5 m) halite interlayer appears within anhydrites. The total thickness of Cisuralian deposits varies very considerably peaking 310 m northwest of the impact structure. Artinskian and Kungurian deposits are lacking in the area due to a regional gap in sedimentation and denudation.

The Guadalupian Series is represented by deposits of all three stages. The Roadian (Kazanian in the Russian Stratigraphic Chart) varies significantly both in vertical section and laterally and by thickness; in some sections southwest of the impact structure, it is absent at all, while northeast of the crater edge, it reaches 74 m thick. It represents there a bioherm massif composed of organogenic (mainly brachyopode), light-grey limestone with dolomite interlayers and gypsum nodules; in places (at the southeast of the crater), quartz sandstone and marl interlayers appear in the base. East of the crater, Guadalupian deposits are represented by dolomite and gypsum with subordinate limestone interlayers in their lower (30 m thick), while the top (20 m) of the sequence is dominated by organogenic limestone with marl interlayers. All listed rocks are overlain by a red continental sequence involving deposits of Wordian, Capitanian, Wuchiapingian, and Changhsingian Stages.

The Wordian Stage (55–85 m thick) includes Nizhnee Ustje and Sukhona Formations. The former one of 33–59 m thick consists of clays with interlayers of siltstone and, more rarely, sandstone, marl, dolomite, and limestone, the contribution of sandstones increasing from northeast to southwest so that along the southwest edge of the Puchezh-Katunki structure, sandstones contribute more than 50% of formation volume. All terrigenous lithologies have carbonate-gypseous cement; gypsum lenses and nodules are common as well. The Sukhona Formation (22–39 m thick) is dominated by marls (at the east) or clays (at the west); it differs by a gradual reduction of gypsum upward in the section.

Capitanian and Lopingian deposits (47–102 m) are distinct from Wordian deposits by the increasing occurrence of carbonates and lacking in gypsum, It consists of interlayered calcareous clays, polymict siltstone, rare sandstone, limestone, dolomite, marl; in places, the latter becomes a predominant lithology.

The pre-impact platform cover sequence is crowned by the Lower Triassic, which does not exceed 100 m thick along the northern edge of the Puchzeh-Katunki impact structure; at the southern of the area, Lower Triassic deposits were absent during the impact event—due to either Triassic to Early Jurassic denudation or gap in sedimentation. The Lower Triassic is represented by both Induan and Olenekian Stages. It consists of variegated terrigenous lithologies, which form from three to six rhythmites. The latter open from fine-grained sandstone and are completed by clays. Upward in the section, the rhythmic structure of the sequence becomes obscure to be replaced gradually by thin alternation of sands, sandstones, siltstones, and predominant clays.

Thus, an averaged lithological composition of the sedimentary cover at the impact site is as follows: carbonates contribute about 45%, dolomites being predominant over limestones (with ratio 2:1); claystones and argillites, about 32%; siltstones and sandstones, 12%, marls, 3%, and anhydrite plus gypsum, 8%.

2.1 Paleoenvironmental Conditions at the Time of Impact

The paleogeographical environment at the time of impact affects significantly many features of an impact structure including crater morphology; formation, amount, and relationships of various impact lithologies; preservation of the impact structure during the subsequent history of the area.

As stated above, the Puchezh-Katunki impact occurred in the two-layered target: the crystalline basement and thick (ca. 2 km) overlying platform cover consisting of terrigenous, carbonate, and evaporite rocks. The main body of the platform cover was accumulated before the end of the Carboniferous. Marine transgressions peaked in the Famenian; after this time, vast intra-continental basins revolved to increase the sedimentary sequence. However, the Rodian age is terminated by the final disappearance of marine environment. Starting from the Wordian age, denudation predominates over sedimentation. Red beds accumulated in a vast fresh-water lake, which disintegrated to some local basins in the Capitanian. The Lopingian epoch is characterized by the arid climate and decrease of sedimentation.

In the Induan, the area under consideration was a drainless depression, where numerous streams brought to the terrigenous material. Within this lowland aligned to the young alpine-type Ural mountains, lacustrine, lagoon, alluvial, and proluvial continental sediments were deposited. Nearly to the late Induan, an inflow of coarse terrigenous material almost ceased. During the Olenekian, the most of the area submerged while streams acted more appreciable. Arid climate changed to humid one, so both sedimentation and biota also changed. However, the Early Triassic is terminated by a regional elevation followed by disappearance of lake basins and ceasing of any sedimentation. Both Middle and Late Triassic as well as Early Jurassic and Aalenian deposits are absent in East European Platform except for its western, southwestern, and southeastern peripheral areas. Ascending tectonic movements culminated in Later Triassic and Early Jurassic to cause the intense denudation. The most of Olenekian and, in places, Induan deposits were removed; southeast of the impact site (at a peripheral part of the Volga-Urals continent), the denudation affected Lopingian deposits as well. Thus, the topography of the area at the time of the impact was smooth due to preceding denudation.

A sedimentation in the Moscow syneclise recommenced in the Middle Jurassic. Local areas of deposits assigned to the Bajocian by palinostratigraphic analysis (Kordun et al. 1975; Golts and Nazhestkina 1983) occur at 100–200 km northwest (Sysola Formation) and southwest (Kudinovo Formation) from Puchezh-Katunki. These are lagoon and partially lacustrine formations of no more than 30 m thick. They consist of interlayered clays and silt, with subordinate sand and shale. In the unified regional stratigraphic scheme of the Jurassic of East European Platform (Mitta et al. 2012) these deposits are juxtaposed with the Kovernino Formation (post-impact crater lake deposits of the Puchezh-Katunki impact structure). Bathonian deposits (Moksha Formation) in the east of Moscow syneclise are also continental (lagoon or lacustrine). This formation of 20–25 m thick is a rhythmic unit consisting of thin interlayered quartz sand, silt, and clay; similar deposits occur in Kovernino depression where they constitute the top of crater lake sediments. A marine basin spread to the Puchezh-Katunki area from the north no earlier than in the Callovian, i.e. a long time after the impact event.

Many features of impact formations: a relatively small amount of impact melt; the sharp predominance of fragmental impact melt rocks (suevite) over massive ones (tagamite); intense alteration of impact glasses; extensive post-impact hydrothermal activity, indicate that the target was rich in water. Thus, it may be assumed that the impact site was a NW inclined accumulative plain with numerous lagoon and lake basins.

References

Bogdanova SV, Lapinskaya TA (1989) Stratigraphy of Lower Precambrian formations of the Volga-Urals Terrain. In: Shurkin KA (ed) Stratigraphy of the USSR. Lower Precambrian. European part of the USSR, vol 2. Nedra Press, Moscow, pp 89–105 (in Russian)

Bogorodskaya OA, Tumanov RR (1980) Geological map of the USSR on 1:200,000 scale. Sheet O-38-XXVI: Explanatory notes. Moscow, 132 p (in Russian)

Bogdanova S, Gorbatschev R, Garetsky RG (2005) EUROPE: East European Craton. Encyclopedia of geology, vol. 2. Elsevier, pp 34–49

Fadeeva LI (ed) (2000) Explanatory notes to State Geological Map of Russian Federation of 1:000 000 scale (new generation). Sheet O-37(38)—Nizhny Novgorod. VSEGEI Press, St.Petersburg, 262 p (in Russian)

Golts SI, Nazhestkina SI (1983) Geological and hydrogeological map of the USSR on 1:200,000 scale. Sheet O-38-XIII: Explanatory notes. Moscow (in Russian)

Kirikov VP (ed) (2002) Legend of the State Geological Map of Russian Federation of 1:1 000 000 (3d generation). Central European series of sheets. SIC "Geocentre", Moscow (in Russian)

Kirkov IG, Kochergina VA, Khaidarova DA, Naumov MV (2019) Geological map of Russia Federation on 1:200,000 scale, 2nd generation. Sheet O-38-XXVI (Puchezh): Explanatory notes. Moscow, 130 p (in press, in Russian)

Kordun BM, Zhuravlev AV, Sangatulina DG (1975) Geological map of the USSR on 1:200,000 scale. Sheet O-38-VII. Explanatory notes. Moscow (in Russian)

Masaitis VL, Pevzner LA (eds) (1999) Deep drilling in the Puchezh-Katunki impact structure. VSEGEI Press, Saint Petersburg, 392 p (in Russian)

Mintz MV (ed) (2010) Deep structure, evolution, and commodities of Early Precambrian basement of East European Platform: Interpretation of seismic data from base profile 1-EB and profiles 4B and Tatseis, vol 1, 408 p; vol 2, 436 p. GEOKART, GEOS Press, Moscow (in Russian)

Mitta VV, Alekseev AS, Shik SM (eds) (2012) Unified regional stratigraphic chart of the Jurassic of East European Platform. Explanatory note. PIN RAS – VNIGNI, Moscow, 64 p (in Russian)

Muslimov RK, Lapinskaya TA (eds) (1996) Crystalline basement of Tatarstan and problems of its oil-and-gas contents. Delta Press, Kazan 486 p (in Russian)

Naumov MV, Larionov AN, Masaitis VL, Mashchak MS, Bogdanova SV, Presnyakov SL, Lepekhina EN (2015) Isotopic age determination of shocked crystalline rocks from the central part of East European Platform (Vorotilovo Deep Drillhole). Reg Geol Metallog 61:79–90 (in Russian)

Tumanov RR, Bogorodskaya OA (1975) Geological map of the USSR on 1:200,000 scale. Sheet O-38-XXVI (in Russian). VSEGEI Press, Leningrad

Ulanov EI, Pisannikova EL, Chumakov OE (2005) Legend of the State Geological Map of Russian Federation of 1:200 000 scale. Middle Volga series of sheets. Nizhny Novgorod, State enterprise "Volgageologia" (in Russian)

Mode of Occurrence and Composition of Impact-Generated and Impact-Modified Formations

Victor L. Masaitis, Mikhail S. Mashchak, Mikhail V. Naumov and Tatjana V. Selivanovskaya

The crater fill includes a series of formations: sedimentary variegated mesobreccia (block-rubble); sedimentary megabreccia; polymict mesobreccia (block-rubble); suevites; polymict microbreccia, and tagamite (massive impact melt rocks). The latter forms either thin veins within deformed rocks of central uplift (authigenic breccia) or irregularly shaped bodies within polymict allogenic breccia on its slopes. The total amount of impact melting products is estimated to not exceed 3.5 km^3. Both tagamites and suevites comprise polycrystalline impact diamonds. The compositional compliance of impactites with target rocks (crystalline with a minor admixture of a sedimentary constituent) is evidenced by various geochemical and isotope data.

1 General Characteristics of the Impact Structure

The Puchezh-Katunki impact structure with a diameter of about 80 km is situated in the basin of the Volga River. It is centered at 56° 58′ N, 43° 43′ E at 75 km from the city of Nizhny Novgorod. The impact structure has not any manifestation in the present topography due to post-impact erosion and accumulation as well as to recent relief-forming processes. The impact site now appears as a denudation-accumulative plain crossed by the valley of Volga River. Absolute hypsometric marks of watersheds vary from 110 to 174 m. In floodplains of Volga and Uzola rivers, they decrease to 74–94 m. Within the Volga River valley, the plain is dissected by deep ravines with rather steep (up to 10°) slopes.

The interior morphostructures of the crater and distribution of impact-derived formations are shown in Figs. 1, 2 and 3. The main structural features of the Puchezh-Katunki as well as spreading and interrelations of its impact-derived formations are

V. L. Masaitis (✉) · M. S. Mashchak · M. V. Naumov · T. V. Selivanovskaya
A.P. Karpinsky Russian Geological Research Institute, 199106 Sredny Prospekt 74, St. Petersburg, Russia
e-mail: vcmsts@mail.ru; Victor_Masaitis@vsegei.ru

V. L Masaitis and M. V Naumov (eds.), *The Puchezh-Katunki Impact Crater*, Impact Studies, https://doi.org/10.1007/978-3-030-32043-0_3

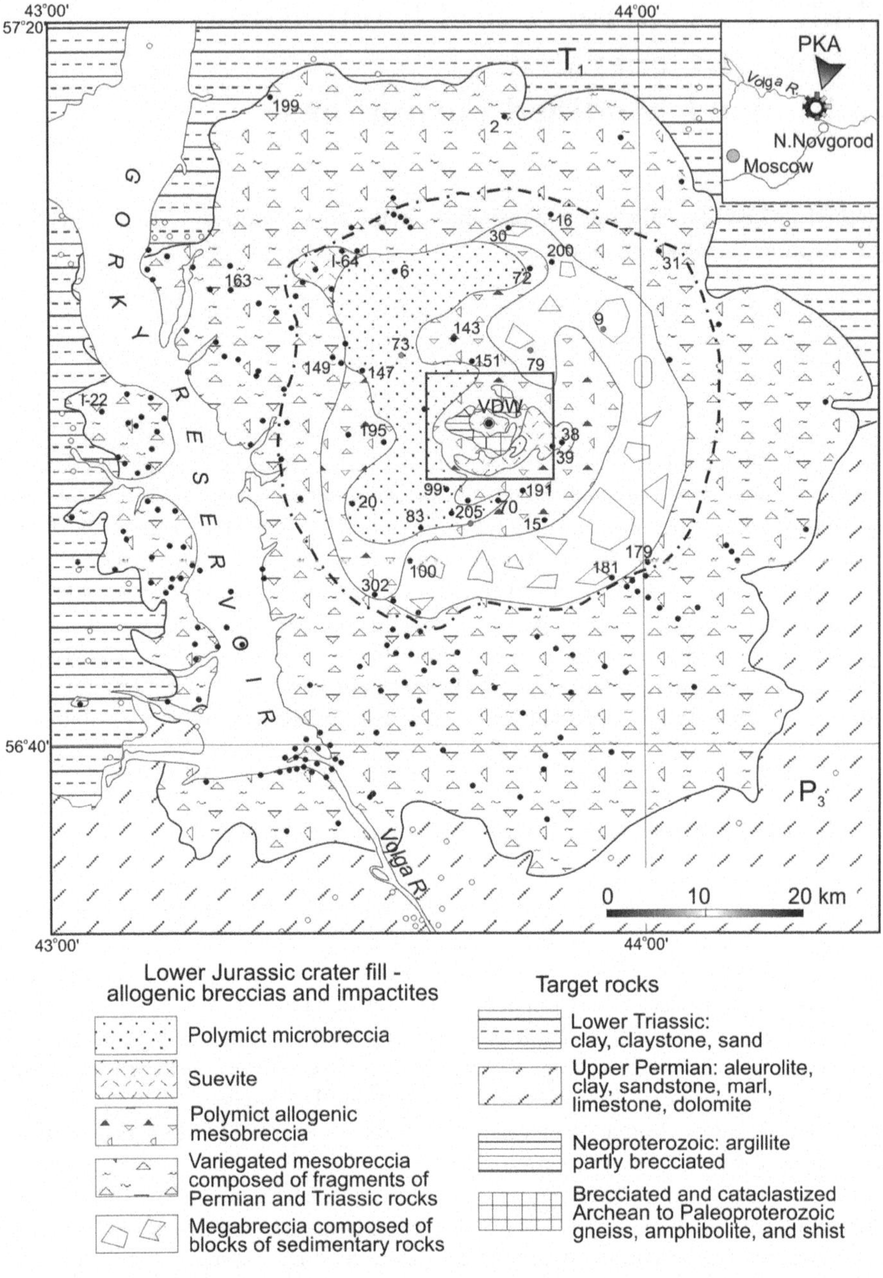

Fig. 1 Geological map of the Puchezh-Katunki impact structure (crater lake deposits and overlaying formations are not shown). Dash-and-dot line, an outer limit of spreading of Jurassic crater lake clays, coptomict sandstones and gritstones. All boreholes penetrated into impact-derived formations are shown by black circles; boreholes penetrated into target rocks, by open circles; boreholes stopped in crater lake deposits, by grey circles. Boreholes referenced in the text are numbered. Contour of the geological map of the central uplift (Fig. 2) is shown by a rectangle

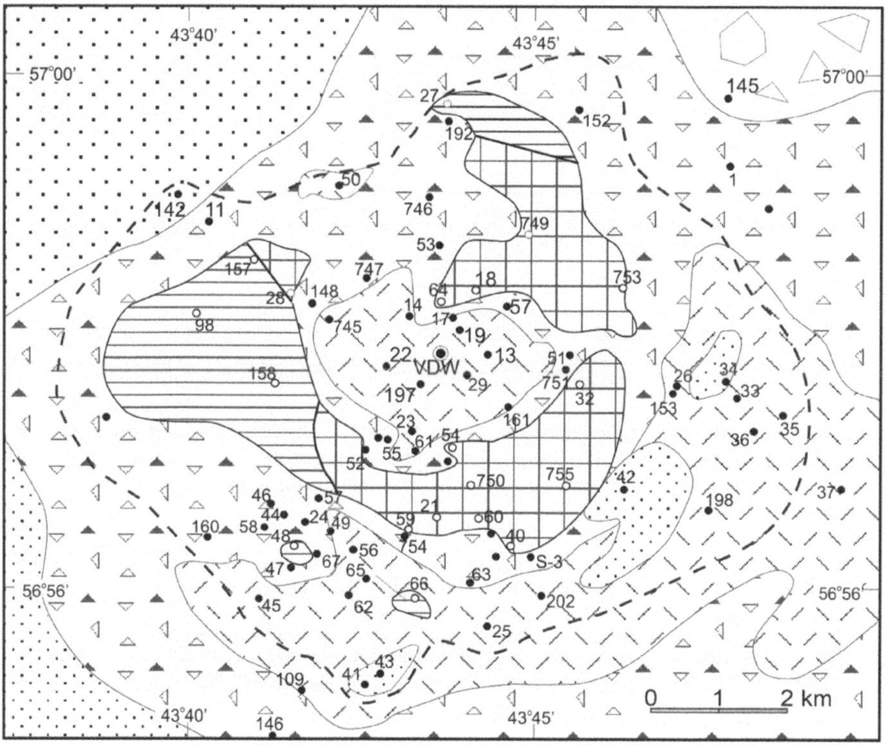

Fig. 2 Geological map of the central uplift of the Puchezh-Katunki impact structure (crater lake deposits and overlaying formations are not shown). For legend, see Fig. 1. Dashed line, an inner limit of megabreccia distribution under overlying formations. All boreholes penetrated into crater fill formations are shown by black circles; boreholes penetrated into deformed target rocks (authigenic breccia), by open circles; VDW, by a double circle. Boreholes referenced in the text are numbered

determined in general (besides the collision of projectile) by a composite structure of the target, which comprises two components: Archean to Paleoproterozoic crystalline basement a and overlying Neoproterozoic to Lower Triassic sedimentary cover. All these formations are described in Chap. 2. The main structural elements of the impact structure are (Fig. 3): (1) ring terrace (outer ring terrace), (2) annular depression (annular trough), and (3) central uplift of crystalline rocks ("Vorotilovo protrusion") with a central pit. In the outer terrace, annular trough, and partly on steep external slopes and in the center of the uplift there are distributed rocks that fill the impact structure—various allochthonous impact-derived formations (crater fill) that overlie authigenic brecciated sedimentary and crystalline rocks (target rocks or bedrocks). The target rocks near the outer boundary of the ring terrace form a low structural rim.

The crater fill includes impact lithic breccias and impactites. The former are composed of blocks and fragments of sedimentary and crystalline rocks, in places with minor admixture of impact glass fragments. Several lithic breccia formations are

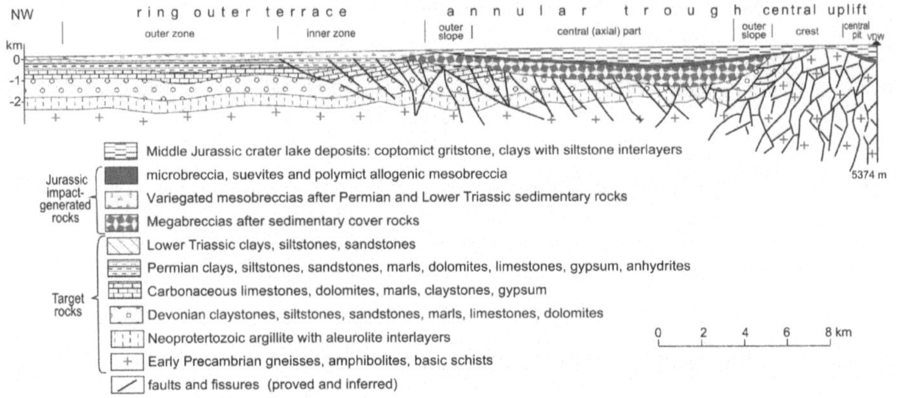

Fig. 3 Geological section of the northwest sector of the Puchezh-Katunki impact structure (overlaying deposits are not shown). All morphostructural elements of the impact structure and VDW location are shown. Outer slope of the central uplift is the inner slope of the annular trough. Vertical and horizontal scales are the same

distinguished, these are: sedimentary variegated mesobreccia (block-rubble), sedimentary megabreccia, polymict mesobreccia (block-rubble), and polymict microbreccia. Impactites are impact-derived lithologies consisting completely or partially of products of impact melting. Chilled impact melt forms irregular bodies or small veins (tagamites), or accumulations of fragments, bombs, and lapilli (suevites). Both impactite varieties contain numerous clasts of target rocks and their minerals.

Comparative characteristics of all distinguished crater fill formations are given in Table 1.

All crater fill units are overlain with unconformity by crater lake deposits (coptomict gritstones and sandstones and clays), and by regionally distributed Mesozoic and Cenozoic sediments.

Detailed descriptions of both structural elements of the crater interior and crater fill are given in the following sections based mainly on abundant drilling data, but results of geophysical and petrophysical surveys contribute as well.

2 Main Results of Geophysical Surveys and Petrophysical Studies

In addition to the extensive drilling, an important information on the inner structure of Puchezh-Katunki crater was obtained from geophysical surveys. Gravimetric and magnetic surveys of variable scale over the area of the impact structure were carried out 1960–1980ths. A review of results of geophysical surveys including seismic prospecting in 1984–1986 and logging data from the VDW, has been given

Table 1 Lithological characteristics and distribution of crater fill formations

Crater fill formation	Composition of clastic material	Prevailing size of fragments (m)	Main area of appearance	S (km^2)	H_{av} (m)	H_{max} (m)	V (km^3)
Sedimentary megabreccia	Neoproterozoic, Devonian, Carboniferous and Permian sedimentary rocks	10n–100n	Annular trough	1050	1090	1500	1150
Variegated breccia	Upper Permian and Lower Triassic sedimentary rocks	0.0n–10n	Ring terrace	3400	100	281	341
Polymict mesobreccia	Basement crystalline rocks, Neoproterozoic and Devonian sedimentary rocks	0.0n–n	Annular trough, central uplift	550	31	170	20
Suevite	Basement crystalline rocks, Neoproterozoic and Devonian sedimentary rocks, impact glasses	0.00n–n	Central uplift, annular trough	80	30	121	2.5
Coptoclastite	Basement crystalline rocks, Neoproterozoic and Devonian sedimentary rocks	0.00n–0.0n	Annular trough, central uplift	350	–	80	–

S—area of distribution
H_{av}, H_{max}—average and maximum thickness
V—preserved volume estimate

by L. A. Pevzner and V. I. Segalovich (in Masaitis and Pevzner 1999). The following description is based on this review. In addition, a lot of information on physical properties of rocks was obtained due to detailed studies of VDW cores and logging carried out by "Nedra" enterprise (G. I. Shamray, O. B. Galkina, A. E. Rosaev, I. V. Neronova, N. N. Usanova and others); the results are presented in Masaitis and Pevzner (1999).

2.1 *Main Geophysical Characteristics*

Seismic data. Some seismic profiles processed at different times cut across the Puchezh-Katunki. The most representative are results of common midpoint shooting (CDP) processing for a seismic profile that cross the northern part of the impact structure. The wavefield patterns are almost symmetric with respect to the crater center structure. Taking into account the circular symmetry of the seismic images, their details are examined for one of the radial intersections (Fig. 4). From the periphery to the center four concentric zones occur that have specific seismic characteristics.

The outer ring zone (from 26 to 34 km off the crater center) is characterized by weak bending of sub-horizontal wave packets represented seismic reflections from stratigraphic boundaries inside sedimentary platform cover (Neoproterozoic to Paleozoic). On the time section, this zone corresponds to the 3–4 km wide and 200 m high ring structural rim. The ring bedrock rim, known from geological data, is projected on it. Most likely, weak deep deformations related to the crater formation are depicted here.

In the upper part of the sedimentary cover (Permian to Triassic), the parallel faults of wave packages are registered. They are inclined towards the center at angles of 70°–80°. In the ring structural rim displacements attenuate at the upper boundary of Carboniferous sediments.

The wave field typical for the most of the sedimentary cover (Neoproterozoic to Upper Paleozoic), is registered within the outer zone in the time interval from 0.22 to 0.28 s. up to 1 s. The greatest reflections in this time interval are very likely correspond to the upper boundary of Permian carbonaceous sediments.

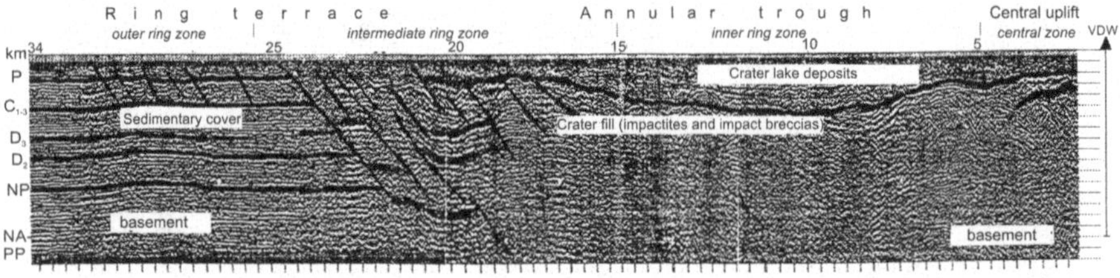

Fig. 4 A section (between boreholes 164 and 192) of the seismic (CDP) line cut across the north-western part of the Puchezh-Katunki impact structure (from Masaitis and Pevzner 1999)

The intermediate ring zone (from 26 to 17 km off the crater center) is characterized by vertical and steeply inclined in echelon faults. The reflecting boundaries between probable faults are distinguished by increasing of deformations towards the center. The amplitude of the latter in the time scale is 150–160 ms; in the depth scale this is 120–140 m. At the same time, fragments of the wavefield between the faults remain parallel indicating that the normal stratification of sedimentary beds remains there.

The inner ring zone (from 5 to 17–23 km off the crater center) is distinguished by the deficit of regular reflected waves starting from 0.55 s. On the contrary, high-intensity, low-frequency reflections appear corresponding to Jurassic sediment formations filling the annular trough. The wave trains recorded in the annular trough zone are clearly bended towards its axis that is located at a distance of 10.0–10.5 km from the crater center. The maximum amplitude of the trough in the time scale is 0.50–0.52 s; this corresponds to a depth of 700–750 m for the bottom of crater lake deposits. The inclination angles of reflecting boundaries are from 5° to 15°.

The central zone has a diameter of about 10 km and is characterized by a discontinuous wave field. As more or less ordered features, axes of distinctly "hyperbolic shape" are recorded that are diffracted. The centers of diffraction belong to the steep slope and the roughness of the central uplift.

The central uplift is crossed by the seismic profile at a distance of 4 km from the geometric center of the impact structure. Thus, the analyzed interval corresponds to chord (but not the diameter) of the central uplift.

The surface of the central uplift in the dome section responds, as a whole, to the train of weakly correlated reflections. It corresponds to thin-layer formations overlapping a highly fractured and shattered rock domain.

Continuous seismic waves at the level of a possible foot of the central uplift (deeper than 2.2 km) are lacking. Thus, the depth of the true bottom of the crater at the foot that was estimated before as −1600 to −1900 m, evidently, may exceed this value.

Gravitational data. The gravity field over Puchezh-Katunki ring structure is presented in Fig. 5. It is determined by the superposition of two effects: (1) the regional field reflecting mostly a structure of crystalline basement, and (2) the effects caused by the impact crater and the deformed rocks under the crater bottom. The approximate density estimates necessary to separate the anomaly related to the impact structure from the regional field are presented in Table 2.

The negative Δg anomaly related to the impact structure is clearly manifested on the images of the observed field (Bouguer reduction) in different scales. It is expressed by abrupt bending of contour lines and by many separate contours. Local deviations of isoanomalies towards maxima of regional highs corresponds to the negative component of the local Δg field, whereas deviations towards minima of regional lows refer to the positive component.

A band of positive Δg values is registered in the southeast of the impact structure. It is located at a distance of 25–30 km from the crater center (Zander et al. 1967).

Within the outer and the intermediate ring zones, Δg contour lines remain almost linear. This indicates that starting from a distance of 25 km from the crater center

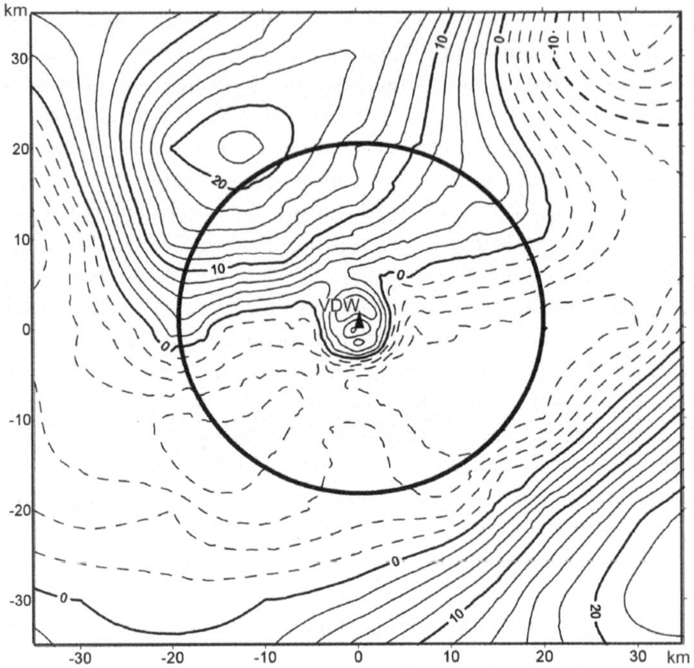

Fig. 5 Observed gravity map (Bouguer reduction) of the Puchezh-Katunki structure (Masaitis and Pevzner 1999). Contour interval, 2 mGal. Contour lines for positive values are solid, for negative values, are dashed. The VDW location in the crater center is shown by a triangle. An approximated contour of a local negative anomaly with a diameter of 42 km corresponding to the crater depression is shown by a solid line. In the crater center there is a positive local anomaly conditioned by the central uplift. It includes central pit and annual crest of the uplift. From south, west, and east the central uplift is surrounded by a ring chain of local gravity lows but from the north, this ring is expressed by bending of isolines towards a regional maximum located north of the crater

outwards, the density of rocks that cause the southeastern anomaly is practically undisturbed by the impact process.

The second band of positive Δg values occupies the northwestern and northern sectors and stretches northward or 80–100 km outward from the crater. In the center, the band is up to 45 km wide. The southeastern part of the anomaly overlaps the central part of the impact structure up to the chord passing 10–12 km southwest of the crater center. The maximum field value (+26 mGal) is registered at a distance of 40 km north-northwest of this line.

An anomaly related directly to the impact was additionally refined by the detailed gravity survey (Fig. 6) in the central part of the ring structure. It was determined that the Δg anomaly achieves the greatest negative value, about -16 to -18 mGal, at a circle line with a radius of 5.5 km. A local positive anomaly of the Δg field is expressed on the negative background in the crest-like relative high located at 2.5–4.5 km from the crater center. At the same time, the Δg values above the crest of the central uplift increase differently in different sectors: in the northern and western

Table 2 Density parameters of main rock formations (after V. I. Segalovich and L. A. Pevzner, in: Masaitis and Pevzner 1999)

Main rock formations	Density parameters		Average thickness (km)	Intensity of Δg anomalies (mGal)
	Average density (g/cm^3)	Redundant density (g/cm^3)		
Crystalline basement				
Leucocratic rocks	2.68	−0.09	40	−10 to −20
Melanocratic rocks	2.90	+0.13	–	+20 to +60
Impact breccias and impactites				
Authigenic crystalline breccia at depths >5 km	2.66–2.88	−0.01 to −0.06	up to 20	−1.0 to −7.0
Authigenic crystalline breccia of the central uplift at depth <3 km	2.25–2.75	−0.05 to −0.25	2.5	−4 to −6
		+0.02 to +0.20		+12 to +16
Allogenic breccias	2.30–2.50	−0.20 to −0.40	2.0–2.2	−8 to −10
Tagamite	2.45			
Suevites	1.88			
Crater lake deposits	2.20–2.30	−0.30 to −0.40	0.03–0.12	0.6–0.8
Platform cover				
NP-PZ	2.50–2.70	−0.02 to −0.12	1.9–3.2	2–10
MZ-KZ	2.10–2.40	−0.20 to −0.40	0–0.5	4–6

For authigenic crystalline breccia of the central uplift at depth <3 km, two estimates of abnormal density parameters and gravity anomalies are given with respect to different environments: in upper line, relative to the crystalline basement, in lower line, relative to crater lake deposits and sedimentary rocks of the platform cover

parts to 10–12 mGal, and in the southern and eastern parts to 12–14 mGal. Moreover Δg values in the annual crest zone of the central uplift are significantly (by 5–6 mGal) lower than its values obtained by interpolation of anomalies between the crater edges.

Geomagnetic data. The magnetic field over the Puchezh-Katunki impact structure is shown in Fig. 7. This map is based on data from the aeromagnetic survey at 300 m height (Tikhomirova 1967). Next works were the surface surveys both at 1:200,000 scale carried out by VSEGEI in 1974, and at 1:50,000 scale headed by E. Pleshakova ("Centergeologija") in 1982. All the data have been adjusted to an unified reference level defined by averaging of the ΔT field. Besides, the data obtained from detailed (1:25,000) ground survey accomplished in 1984 under supervision of M.

Fig. 6 Observed gravity map (Bouguer reduction) of the central part of Puchezh-Katunki structure (after L. A. Pevzner and V. I. Segalovich, in: Masaitis and Pevzner 1999). Contour interval, 1 mGal. Dashed-dot line outlines the outer edge of the annular trough, dotted line, the contour of the central uplift

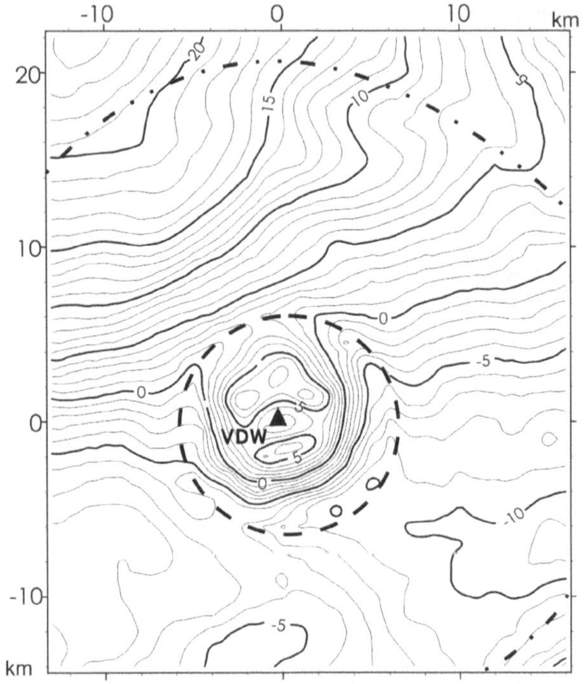

N. Kolesov in the center of the impact structure were taken into account. Previously they were adjusted to the level of the normal geomagnetic field, which differs from the reference level by −400 to −450 nT (Masaitis and Pevzner 1999).

From aeromagnetic data, the Puchezh-Katunki circular structure is marked by a magnetic low of up to 1500 nT amplitude compared to WSW-ENE trending high-amplitude regional anomalies, which most likely are related to basement structure. Although the magnetic low over Puchezh-Katunki shows some internal structure, the latter is characterized by low gradients and amplitudes. No anomalies related to crater fill lithologies are apparent within the impact structure.

The regional field outside the impact structure is dominated by two large sublinear zones of predominantly increased (positive) values and a linear elongated zone of mainly decreased (negative) values that separates the former. The band of negative values crosses Puchezh-Katunki structure in SW-NE direction. On the whole, distribution of positive and negative values of ΔT is similar to gravity anomalies distribution and should be considered in the same way.

At a distance of 20–25 km from the crater center, the isodynamic lines become subconcentric to cut local positive ΔT_a anomalies observed over the ring terrace. Thus, a line dividing regional magnetic anomalies from anomalies related to the impact structure generally corresponds to the crater rim (outer edge of the annular trough). Compared to the surrounding region, the ΔT field in the Puchezh-Katunki is very low (from −300 to −700 nT). However, in the crater center, a high-amplitude maximum

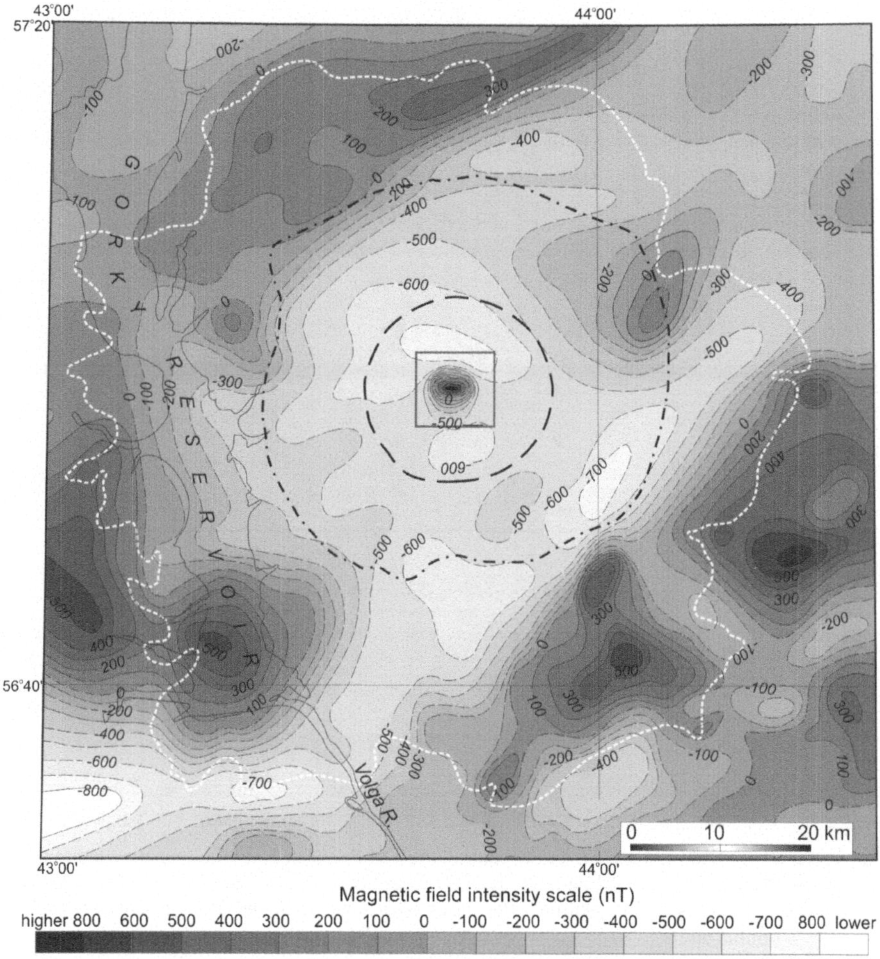

Fig. 7 Aeromagnetic anomalies map over the Puchezh-Katunki impact structure (from Tikhomirova 1967). Isogam interval, 100 nT. Isodynamic lines for positive values are shown by solid lines, for negative values, by dashed lines. Structural elements of the impact structure are shown by thick dashed line (axis of the annular trough), by back dot-dashed line (outer edge of the annular trough), and by white dotted line (outer limit of impact breccias extension). Contour of the magnetic field map of the central uplift (Fig. 8) is shown by a rectangle

with a radius of about 2.5 km occurs (Fig. 7). There, the value of aeromagnetic anomaly increases up to +800 nT and, according to ground survey results, it peaks +2200 nT. From the north, it is limited by an arch-like band of 10 km wide in average, with negative values of up to −750 nT. The central anomaly represents a mating of a crest-like increment (of up to 1920–1980 nT) and bow-shaped band of an increased field (Fig. 8). Evidently, both elements are conditioned by the same source.

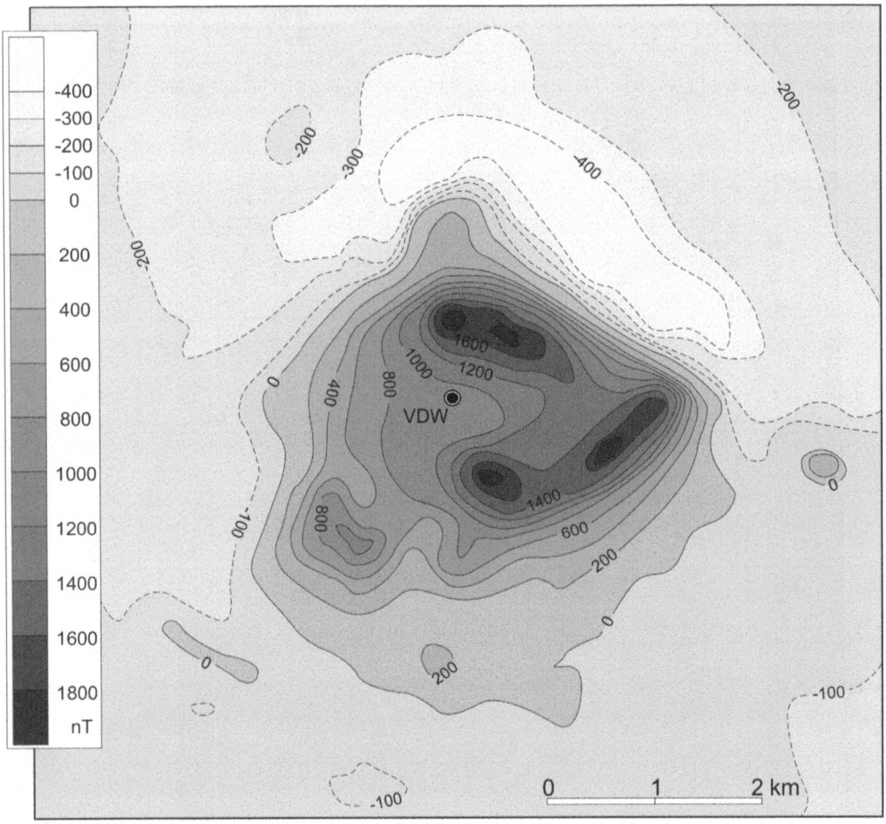

Fig. 8 Reference magnetic field over the central uplift of Puchezh-Katunki impact structure (from Masaitis and Pevzner 1999, modified). Isodynamic lines for positive values are shown by solid lines (contour interval, 200 nT), for negative values, by dashed lines (contour interval, 100 nT). VDW is shown by a double circle

The given data on abnormal magnetic field allow us to consider that its source inside the impact structure is isolated relative to inhomogeneities of the magnetic field of the surrounding region. It might be accepted that magnetization of the central source was newly created, while the magnetic field of Precambrian basement rocks was completely destroyed by the impact.

When the magnetization of rocks is compared with estimated post-shock temperature and pressure for the same rocks, one might find that the boundary separating the magnetic core from the non-magnetic background corresponds to residual post-shock temperatures about 600 °C and shock pressures of 20–25 GPa. The comparison of magnetization and zoning of shock metamorphism of rocks in the VDW core shows that from top to bottom, against a parabolic drop of shock pressure from 40–45 GPa to 30 GPa and a further linear drop from 30 to 15 GPa, there are two great high-temperature zones with maxima of 950–1100 °C at depths of 900–1600 m and 2400 to 3300 m (see Sects. 4.1 and 4.2 for more detail). The temperature maxima are correspond to the highest magnetization values. Within these intervals, the latter have average values of 8.2 and 12.6 A/m.

To summarize geomagnetic data on the Puchezh-Katunki, L. A. Pevzner and V. I. Segalovich (Masaitis and Pevzner 1999) conclude that the magnetic field of this circular structure is extraordinary anomalous for East European Platform. Within a radius of 2.5 km from the crater center, a complete interruption of the magnetic field structure related to Precambrian basement occurs. The magnetic features originated by the impact include a local central anomaly of very high intensity with a diameter of 2.5 km. The magnetic phenomena related to the impact structure may be explained by destruction of the pre-impact ferromagnetism at the subdomain level in rocks under shock pressure above 3–4 GPa. Such a pressure level may correspond to a critical threshold value similarly to the Curie point. The critical threshold is likely to correspond to a change of mineral magnetism carriers into the stable paramagnetic or, rather, superparamagnetic state. Additional cause is the generation of a new high magnetization within the zone where the most high-temperature transformations occurred due to shock and friction heating (close to fusion) during compression, excavation and early modification stages (see Sect. 4.2).

2.2 *Petrophysics of VDW Core*

Physical properties of rocks of the central uplift were measured in VDW cores by different methods and described in detail by I. V. Neronova, L. A. Pevzner, O. B. Galkina and others (in Masaitis and Pevzner 1999). Variation trends of primary physical properties of rocks throughout the VDW section averaged for 50 m thick depth intervals are presented in Fig. 9. Downward in the section, the bulk density increases (from $\sigma_{\min} = 1.84$ to $\sigma_{max} = 2.89$ g cm^{-3}) as well as velocity of transversal waves (from $V_{S_{MIN}} = 1.68$ to $V_{S_{MAX}} = 3.04$ km s^{-1}) and velocity of longitudinal waves (from $V_{P_{MIN}} = 3.93$ to $V_{P_{MAX}} = 5.94$ km s^{-1}), while the open porosity decreases (from $K_{P_{MAX}} = 30$ to $K_{P_{MIN}} = 1$ %). Distribution of magnetic properties of rocks is characterized by an alternation of intervals sharply differing by their magnetization, but in general, values of magnetic susceptibility and natural remanent magnetization tend to decrease with depth. Such an alternation is recorded mostly in the upper part of the authigenic breccia (down to a depth of about 3300 m in the VDW); downward in the section the variability of magnetic properties is much less. Throughout the section, a significant anisotropy of magnetic and elastic properties of rocks is recorded; it is strongly pronounced above 1500 m, and it decreases downward in the section.

Values of physical properties in selected petrophysical intervals represented in Fig. 9, do not obey the Gaussian distribution. This may be somewhat explained by the different composition of rocks within selected depth intervals. However, the variability of magnetic and elastic properties for uniform petrographic samples (e.g., for

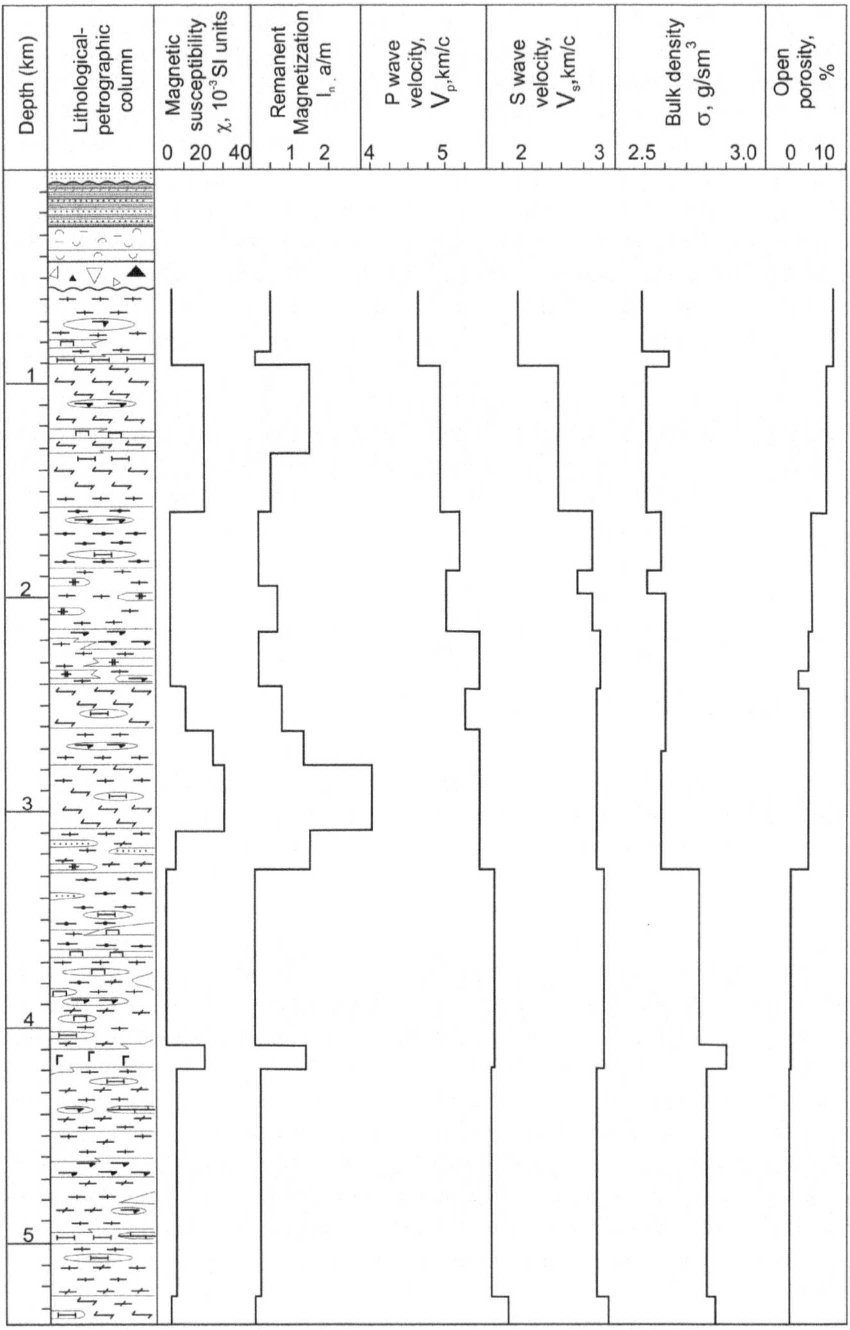

Fig. 9 Sketch petrophysical section of the VDW (after L. A. Pevzner, O. V. Galkina, A. E. Rozaev, N. N. Usanova, in: Masaitis and Pevzner 1999). Physical properties of rocks are averaged for 50 m thick depth intervals. For lithological legend, see Fig. 2 in Chap. 2

gneisses, schists and amphibolites) does not correspond to normal distribution law as well. Thus, a high dispersion of petrophysical parameters in VDW cores is due not only to initial heterogeneity of rock composition, but also due to variations in the level of impact-induced transformations, rock texture features and superposed alteration. Physical properties of tagamites and suevites vary throughout the VDW section. For tagamites, both bulk density and longitudinal wave velocity increases, while the porosity decreases downward in the section. Generally, physical properties of gneisses, amphibolites and schists affected by impact transformations differ considerably from those of Precambrian crystalline rocks composing the basement of East European Platform outside of the impact structure.

3 Interior Morphostructures

As noted above, three main structural elements of the impact structure are distinguished based on drilling and geophysics, these are outer ring terrace, annular trough, and central uplift (Figs. 3, 10 and 11), each of them is characterized by distinctive features of composition, thickness, and distribution of crater fill and crater lake formations. Vertical geological sections typical for these structural elements are given in Fig. 12.

3.1 Outer Ring Terrace

The contours of the ring outer terrace are currently defined by the spreading of allogenic sedimentary variegated breccia consisting mostly of fragments of Permian and Lower Triassic clayey red rocks (Fig. 1). Its width varies from 8 to 12 km in the northeast to 20–30 km in the southwest. The base of variegated mesobreccia is a relatively flat and gently dipping (inclination is no more than 1°–3°) toward the center of the crater. In the northern part of the impact structure, the basement of the terrace is built of deformed rocks of Lower Triassic (Induan Stage) and Lopingian Permian. In the southeast, it is made by rocks of the Guadelupian and Lopingian Series. In some places at the southeastern part of the crater, the rocks of the Cisuralian series are outcropped in the ring terrace in addition. The boundary of the terrace and the annual trough runs approximately along the contour line of −100 m of the true crater floor (Fig. 10). Seismic profiling showed that in Paleozoic strata of the terrace basement a number of structural ring swells with an amplitude up to several tens of meters fading in the radial direction might be outlined.

The structure of the ring terrace is best studied in the western part of the impact structure where the variegated breccia is penetrated by more than 50 holes (Fig. 1) and exposed in numerous cliffs at both western and, partially, eastern banks of the Gorky reservoir, from the Sokolskoe settlement to the mouth of the Trotsa River (Fig. 13). The basement of the ring terrace is complicated there by radial troughs. The

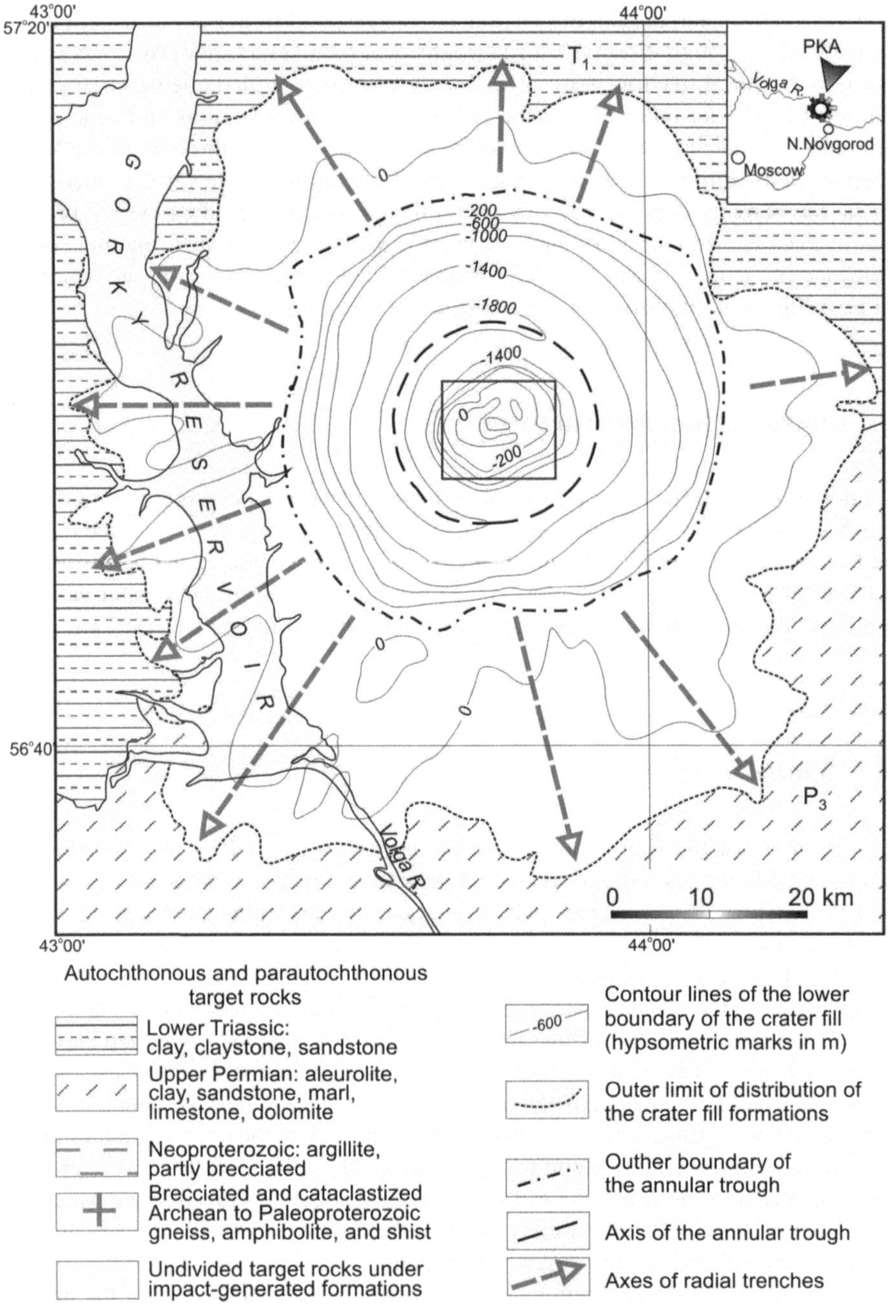

Fig. 10 Geological map of the base of crater fill formations of the Puchezh-Katunki impact structure. The hypsometric position of this surface is shown based on drilling and seismic reflection data. The contour of the geological map of target rocks composing the central uplift (Fig. 15) is shown by a rectangle

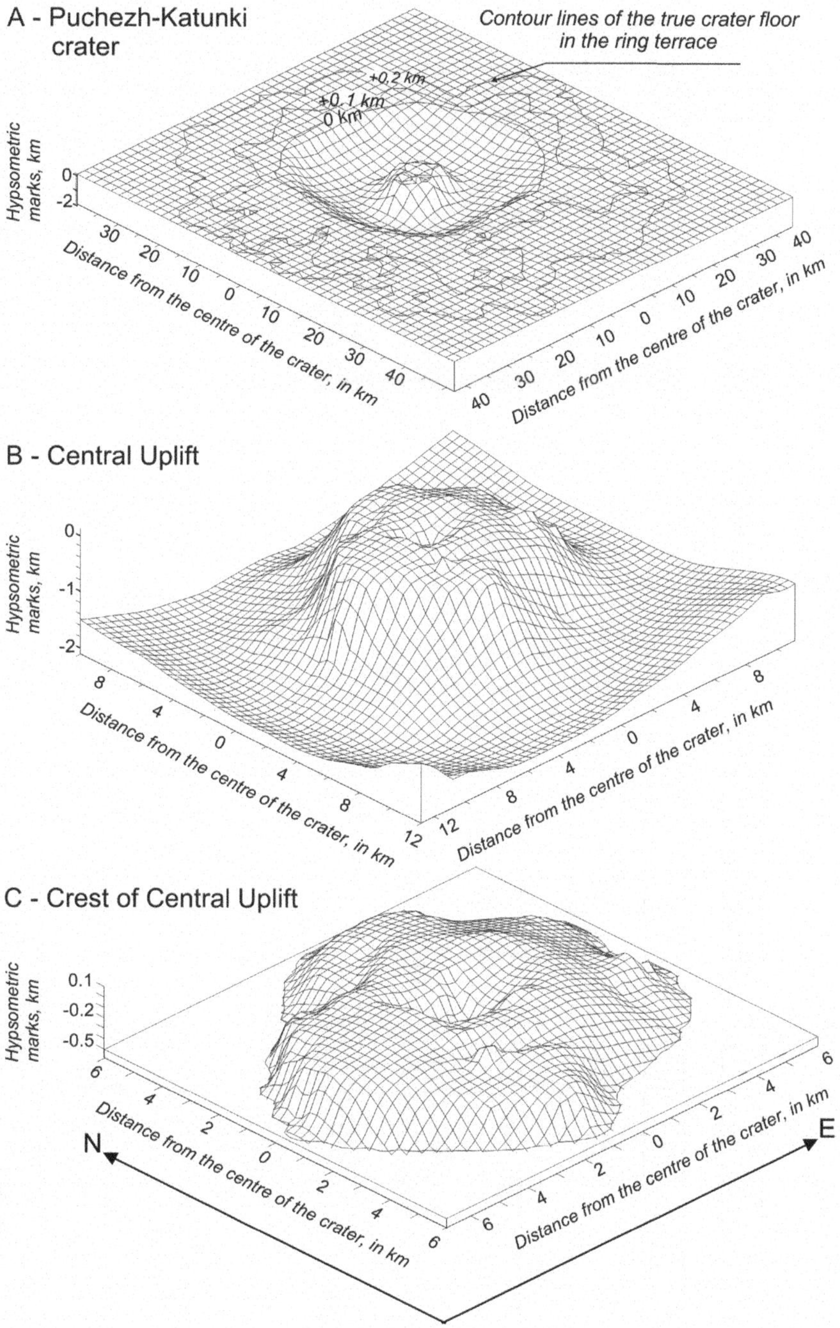

Fig. 11 Three-dimensional map of relief of the lower surface of the crater fill (true bottom floor) of the Puchezh-Katunki impact structure. The entire impact structure, its central uplift, and crest part of the latter are shown with scaling up. Vertical exaggeration is 3:1

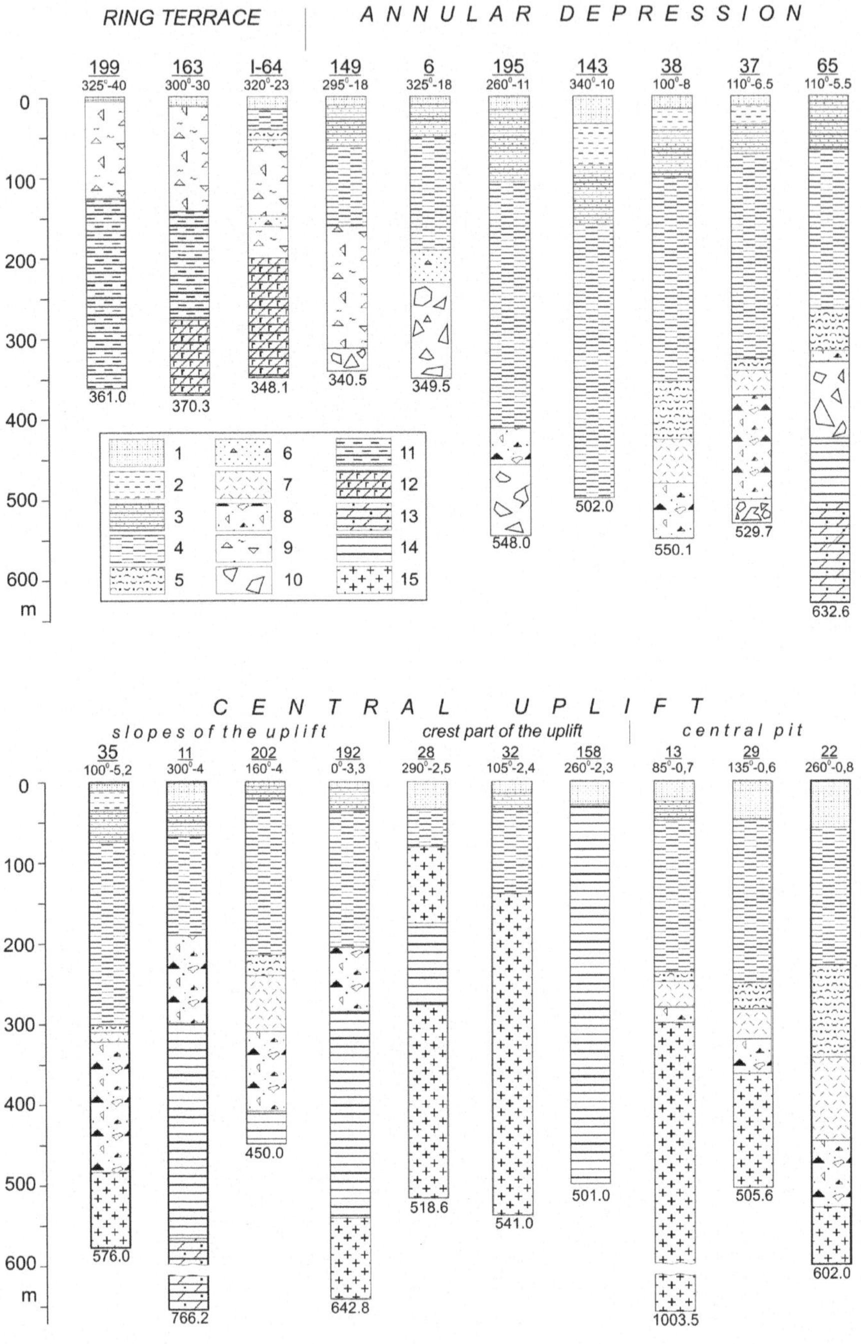
RING TERRACE
A N N U L A R D E P R E S S I O N
199
325°-40
163
300°-30
I-64
320°-23
149
295°-18
6
325°-18
195
260°-11
143
340°-10
38
100°-8
37
110°-6.5
65
110°-5.5
0
100
200
300
400
500
600
m
361.0
370.3
348.1
340.5
349.5
548.0
502.0
550.1
529.7
632.6
1
2
3
4
5
6
7
8
9
10
11
12
13
14
15
C E N T R A L U P L I F T
slopes of the uplift
crest part of the uplift
central pit
35
100°-5,2
11
300°-4
202
160°-4
192
0°-3,3
28
290°-2,5
32
105°-2,4
158
260°-2,3
13
85°-0,7
29
135°-0,6
22
260°-0,8
576.0
766.2
450.0
642.8
518.6
541.0
501.0
1003.5
505.6
602.0

◀**Fig. 12** Generalized geological sections of characteristic boreholes drilled in ring terrace, annular trough, and central uplift of the Puchezh-Katunki impact structure. 1–3, Overlaying deposits: 1, Neogene-Quaternary loams, sands, clays; 2, Lower Cretaceous clays with sand and sandstone interlayers; 3, Middle to Upper Jurassic sands, clays. 4 and 5, Lower to Middle Jurassic crater lake deposits: 4, clays with siltstone interlayers (Kovernino Formation), 5, coptomict gritstones, sandstones, and conglomerates. 6–10, Crater fill formations: 6, polymict microbreccias, 7, suevite, 8, polymict mesobreccia, 9, sedimentary variegated mesobreccia, 10, sedimentary megabreccias. 11–15, target rocks (including brecciated and cataclased ones): 11, Lopingian Permian: clays, siltstones, sandstones, marls, dolomites; 12, Guadalupian to Cisuralian Permian: limestones, dolomites, anhydrites, gypsum; 13, Middle Devonian claystones, siltstones, sandstones, marls; 14, Neoproterozoic argillite with siltstone interlayers; 15, Neoarchean to Paleoproterozoic gneisses, amphibolites, partially granitized basic schists. Permian target rocks are slightly disturbed; Middle Devonian and Neoproterozoic rocks are disturbed and broken into non-unidirectional blocks, often limited by brecciated zones; Neoarchean to Paleoproterozoic rocks are brecciated and cataclased, with numerous mylonite and pseudotachylite veins. Borehole information: in numerator, borehole number; in denominator, azimuth and distance (in km) from the VDW drilled in the centre of the impact structure. Under the column, borehole depth (in m)

latter are distinguished by changes in composition and inner structure of variegated mesobreccias, by increasing of their thickness, and deepening of the base due to excavation by ejected allogenic material. The contact between deformed rocks of the basement and overlying displaced breccia is indistinct. Numerous small irregular folds, thrusts, faults are observed in the transition zone (Fig. 14). This is, more probably, a result of dragging on the surface of ejected material composed of water-saturated plastic sediments.

In the western sector of the impact structure there are four such troughs, which can be traced from the outer slope of the annular trough up to the outer boundary of the area of mesobreccia distribution (Fig. 13). The radial troughs reach 15–20 km long, while their width in frontal parts is from 1.5 to 6.0 km. Their cross-section profiles are trough-shaped with relatively steep (up to 6°–8°) slopes. In the western part of the impact structure, the Puchezh (of western 270° direction) and Nogaevo (of southwestern 250° direction) radial troughs are most clearly expressed (Masaitis and Naumov 1990). The developing of radial troughs in the basement of the ring terrace led to festoon-like outlines of the external boundary of the distribution field of allogenic variegated mesobreccias. The similar lobate contours noted in other sectors of the structure indicate that the radial troughs are likely to exist everywhere in the ring terrace.

The variegated blocky-rubble mesobreccias on the terrace have an average thickness of about 100 m. Within the radial troughs, the thickness reaches 180–200 m, while on the ridges separating them it decreases to the first tens of meters. Variegated breccias comprise abundant large klippen and blocks (in places peaking as much as 3 km^2 in plan, but commonly up to 100–200 m across) of Permian and, much more rarely, Carboniferous carbonate rocks. These klippen and blocks distribute irregularly, but they occur mostly close to the top of variegated mesobreccia. Many of them might be regarded as remnants of a sedimentary megabreccia blanket, which is opened by drilling in the annular trough but is assumed to be removed in the

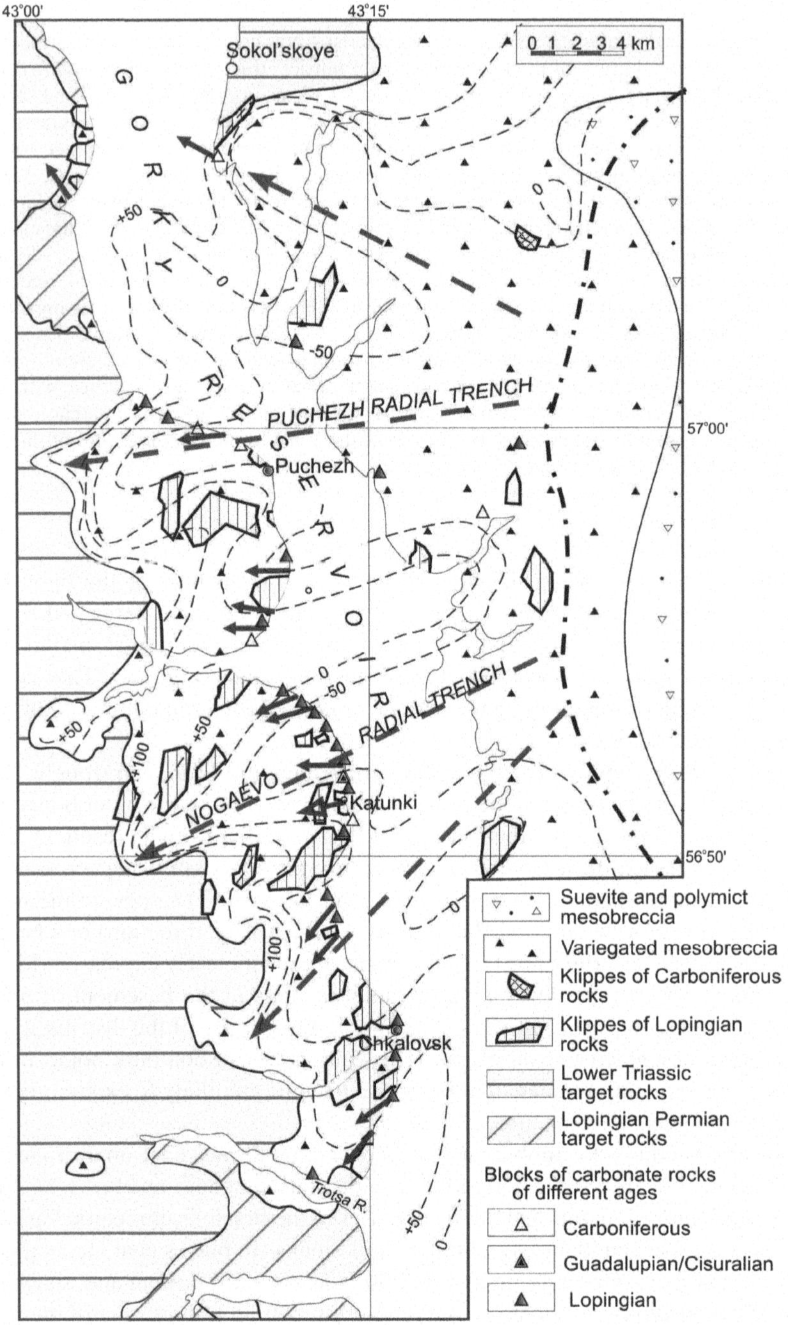

43°00'
43°15'
0 1 2 3 4 km
Sokol'skoye
GORKY RESERVOIR
PUCHEZH RADIAL TRENCH
Puchezh
NOGAEVO RADIAL TRENCH
Katunki
Chkalovsk
Trotsa R.
57°00'
56°50'
+50
0
-50
+100
Suevite and polymict mesobreccia
Variegated mesobreccia
Klippes of Carboniferous rocks
Klippes of Lopingian rocks
Lower Triassic target rocks
Lopingian Permian target rocks
Blocks of carbonate rocks of different ages
Carboniferous
Guadalupian/Cisuralian
Lopingian

◂**Fig. 13** Sketch geological map of the western sector of the ring terrace. Dot-dashed line shows the edge of the annular trough. Thick short solid arrows indicate direction of lineation in clayey blocks in variegated mesobreccia. Axes of radial trenches are shown by thick dashed lines with arrow. Thin dashed lines are contour lines of the base of variegated mesobreccias based on drilling data. Contour interval, 50 m. Boreholes are shown by black circles

Fig. 14 Folded Lopingian clayey red rocks lying at the base of variegated mesobreccia in the western sector of the ring terrace. The western bank of Gorky Reservoir, near Chkalovsk

ring terrace. Commonly, variegated mesobreccia is overlain only by a thin (no more than 30 m thick) sequence of Neogene and Quaternary sands and loams (Fig. 12, boreholes 163, 199).

3.2 *Annular Trough*

Inferred from gravimetric and seismic refraction data (see Sect. 2) as well as the results of drilling (N. D. Fokshansky, V. Ya. Gwin, A. G. Averbukh, O. S. Bogatyrev etc., pers. comm.), the annular trough has an outer diameter of 40–42 km and depth ca. 1.9–1.6 km (Fig. 3). It is filled with various allochtonous impact-derived rock units overlain by crater lake deposits and more recent sediments (Figs. 3 and 12). The underlying target rocks are opened by wells only in the outer and inner slopes of the trough.

The outer slope of the annular trough where the latter joints to the ring terrace is a relatively steep (from 8° to 12° in the upper part of the slope to 20°–25° at the base). The inner slope (outer slope of the central uplift) is also steep. The outer slope is built of layered Permian, Carboniferous and Upper Devonian rocks. They are radially and concentrically fractured as flake-shaped lumps and blocks, which are displaced relative to each other by up to several hundreds of meters. Small subsided blocks of the upper part of the outer slope are overlaid with remnants of variegated mesobreccia. Inward the annular trough, a brecciated Paleozoic stratum of the outer slope transits into sedimentary megabreccia, which lies at the base of the trough. This transition zone is conditionally taken as the bottom of the annular trough, but it does not reflect clearly on seismic profiles.

This parautochtonous basement of allogenic formations is traced by boreholes 1, 35, 11, 202, 192 (Fig. 12) at the inner slope of the annular trough bordered to the central uplift. A thick sequence of large brecciated blocks of sedimentary and crystalline rocks showing features of inverted stratigraphy occurs there beneath of the sedimentary allogenic megabreccia. Some crystalline blocks with veins and irregular bodies of chilled impact melt may be positioned there due to sliding from the central uplift (bh. 192).

According to seismic data, an axial diameter of the annular trough is about 20–22 km (Fig. 11), the base of allogenic formations in the central part of the trough is relatively flat (Fig. 3). The relief of the base is complicated by arc ridges and irregularly shaped elevations with heights up to several tens meters, but in places, even up to hundreds of meters. Certain ridges apparently extend probably for some kilometers, and they alternate in echeloned folds. Nearly concentrically aligned relative to the crater center ridges are predominant.

In the north and the west, the surface of the trough true floor slopes brecciated and slightly displaced layers of Permian, Carboniferous, Devonian and Neoproterozoic rocks, while in the south and the east, Permian, Carboniferous, and Devonian beds (Fig. 3). The intensity of deformation attenuates toward the crystalline basement and in radial direction.

The roof of the crystalline basement is traced from gravity data to be slightly undulating due to alternation of discontinuous concentric, arcuate or linear elevations and lows. In general, three zones of concentrically arranged depressions can be distinguished with radii of 6–7, 10–11, and 16–18 km. The former zone located at the foot of the central uplift is best expressed; the basement roof is sinking there down to 2.6 km deep.

Allochtonous units in the lower part of annual trough are represented by sedimentary megabreccia formed by mixed and displaced klippen and blocks of Neoproterozoic, Devonian, Carboniferous, Cisuralian and Guadalupian formations (Fig. 1). The maximum thickness of this unit in the axial part of the annual trough in its western sector is probably 1.5 km, but in its eastern sector, it does not exceed 0.8–1.0 km. These figures might be overestimated owing to uncertainty of the depth of the lower boundary of allogenic megabreccia.

At the inner slope of the trough and its axial part as well as in the northwestern and southwestern sectors of the outer slope, the megabreccia is successively covered by polymict mesobreccia, in some places by suevites and polymict microbreccia (Figs. 1 and 12, bhs. 1, 37, 195). At the edges of the trough, polymict microbreccia lies directly upon sedimentary megabreccia (Fig. 12, bh. 6). At the outer slope in the eastern of the crater, sedimentary megabreccia is overlain by variegated mesobreccia (Fig. 1). In some places on the slopes of the annular trough and in its axial part, polymict mesobreccias, but mainly suevites and microbreccia, are washed out and transit gradually into coptomict gritstones and sandstones of crater lake deposits (Fig. 12, bh. 1, 37, 38). The latter reach 100 m thick in places. In general, polymict mesobreccia, suevites, and polymict microbreccia form a flattened lens, the thickness of which may peak 300 m (Fig. 3). Its center is shifted relative to the crater center to the northwest. In the western sector of the crater, the rocks forming the lens extend within a radius up to 22 km, while in the east, from 8 to 15 km.

The total volume of polymict mesobreccia in the western sector of annular trough is 378 km^3, while in the eastern sector, only 158 km^3; corresponding suevite volumes are 51 and 31 km^3 (Table 3). It should be noted in addition, the thickest bodies of consolidated impact melt are located within polymict breccia on the southwestern slope of the central uplift.

All allogenic lithologies and coarse-grained crater lake sediments are covered by flat-bedded bedded clays of the Middle Jurassic Kovernino Formation with a maximum drilled thickness of 340 m (Fig. 12, bh. 143). The latter are covered in turn by overlapping Mesozoic and Cenozoic formations with a total thickness of up to 162 m (Fig. 12, bh. 143).

3.3 Central Uplift

The central uplift is the most pronounced and well studied morphostructural element of the Puchezh-Katunki crater due to extensive drilling; apart from the VDW, about 80 boreholes were drilled there, 20 of them are more than 500 m deep (Fig. 2). The central uplift is built of brecciated crystalline rocks of the Neoarchean to Paleoproterozoic, which are overlaid by deformed Neoproterosoic and Middle Devonian sedimentary rocks at its slopes. In addition, blocks of the Neoproterozoic make up the western part of the crest of the central uplift. In the central part of the uplift, crystalline basement rocks form a core up to 7 km across, which is displaced eastward relative to the uplift center. The northern, eastern and southeastern borders of this core coincide approximately with a contour line of the uplift roof −200 m, while its western and southwestern borders cross through the crest and intersect contour lines from 0 to −200 m (Fig. 15).

The morphology and interior structure of the central uplift are shown in Figs. 11 and 15. In terms of the plan, it somewhat elongates in the latitudinal direction in accordance with the prevailing strike of crystalline basement rocks. The uplift is 8 × 10 km in the crest, 12 × 14 km at the base. The steepness of the slopes ranges from 7° to 12° in the arched part to 40°–45° at the base. The slopes are often stepped and

Table 3 Estimates of suevite and polymict allogenic breccias extent preserved in the Puchezh-Katunki impact crater

Structural element	Parameter	Polymict allogenic breccias					Suevites				
		Sector of the impact structure					Sector of the impact structure				
		North-east	South-east	South-west	North-west	Total	North-east	South-east	South-west	North-west	Total
Central pit	*S*	2.7	2.3	1.7	2.8	9.4	1	1.9	1.4	1.4	5.8
	h	34	31	36	53	39	22	27	37	43	33
	V	0.09	0.07	0.06	0.15	0.37	0.025	0.055	0.05	0.06	0.19
Slopes of the central uplift	*S*	15.6	16.3	20.7	14.1	66.7	4.5	14.1	7.8	0.3	26.4
	h	28	42	28	52	37	34	51	20	36	40
	V	0.44	0.69	0.58	0.73	2.44	0.16	0.72	1.16	0.01	1.05
Annular trough	*S*	97.3	69.2	178.9	166.3	511.7	–	9.8	–	40.3	50.1
	h	36	25	30	38	31	–	29	–	23	24
	V	3.5	1.73	5.31	6.32	16.86	–	0.28	–	0.94	1.22
Impact structure as a whole	*S*	69.7	87.9	201.3	177.1	592	5.5	25.8	9.2	42	82.6
	h	18	28	30	39	31	33	41	23	24	30
	V	1.28	2.49	5.95	6.97	19.67	0.185	1.055	0.21	1.01	2.46

S—area of distribution (km^2); *h*—average thickness (m); *V*—volume (km^3)
Certain sectors are conditionally supposed according to direction from the crater center

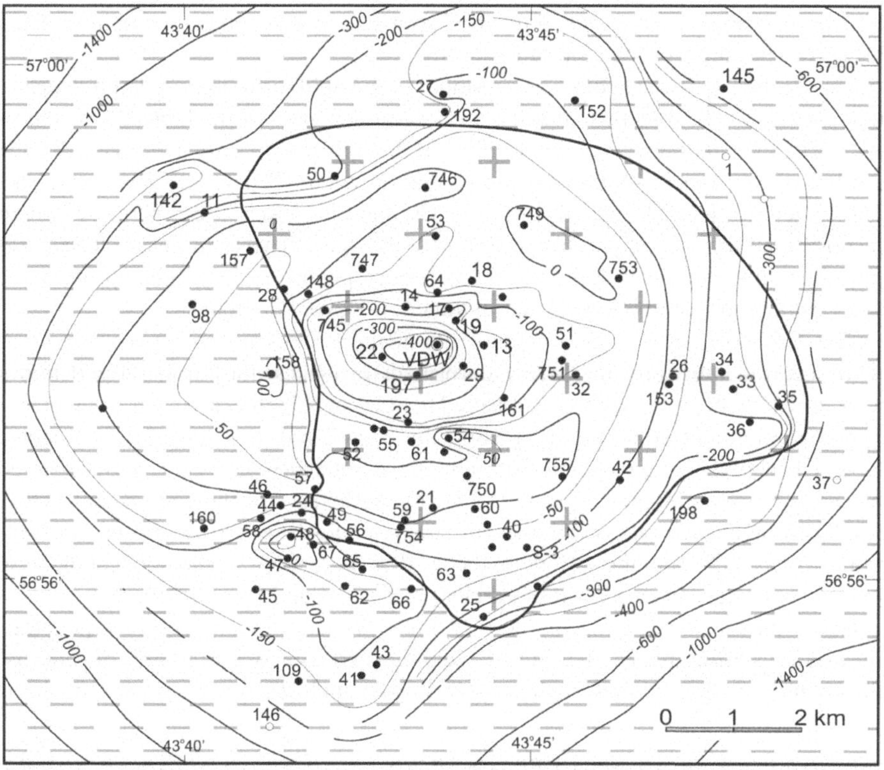

Fig. 15 Relief map of the central uplift. Contour lines of the surface of authigenic breccia are shown based on drilling data. For legend, see Fig. 10. Boreholes penetrating into authigenic breccia are black circles; boreholes stopped within the crater fill, are open circles

precipitous. The relief of the most elevated part of the uplift is uneven; it represents a discontinuous ring crest with hypsometric marks from −50 to +105 m (Fig. 15).

The rocks composing the central uplift occur as blocks ranging from a few meters to several hundred meters across. They are intensely deformed and shock-metamorphosed. Inferred from significant differences in composition, rock age, level of shock metamorphism between nearby blocks, the latter are frequently shifted relative to each other to a considerable extent on the slopes of the uplift. Block shifts are especially clearly recorded in boreholes drilled close to boundaries between blocks of crystalline and sedimentary rocks (Fig. 12, bhs. 1, 11, 28, 192). For example, 2.5 km northwesterly from the uplift center (bh. 28), a 100 m thick sequence of brecciated Neoproterozoic argillite resting on crystalline rocks was opened under another block of intensively brecciated crystalline rocks, which extends for 96 m in core. In the borehole 11, 4 km north-west of the uplift center, Middle Devonian terrigenous rocks were found beneath of a 265 m thick block of Neoproterozoic argillite. As it stated in the previous section, the accumulations of large blocks may be considered as parautochthon. The patterns of inverted stratigraphy can appear due to imbricated thrusts or slipping from the top of the uplift.

The southeastern and southwestern slopes of the uplift are composed of brecciated dark grey and reddish-brown argillite with siltstone lenses of the lower Neoproterozoic formation. In the west and northwest of the uplift, light grey and greenish-grey argillite of the upper Neoproterozoic formation occur. Blocks of Devonian rocks represented by greenish-gray argillites, marls, and limestones, are penetrated only in the northwest (bh. 11) and northeast (bh. 1) of the uplift. Thus, the west of the central uplift including both its slopes and crest is formed predominantly of Neoproterozoic rocks, while the eastern slope is mainly made of Devonian blocks.

The Neoproterozoic and Devonian sedimentary rocks occurring on the slopes of the central uplift are intensively deformed; in places, they transformed into a boulder-rubble breccia, which is subjected to coherent displacements in some places (parautochthon). In some cases, it might be considered as a monomict allogenic breccia. Fragments of terrigenous lithologies are found also in breccia veins injected into the gneisses at depths of up to 0.5 km from the roof of the central uplift (bhs. S-3, VDW, and some others). Numerous shatter cones (Fig. 16) were found in Neoproterozoic argillites in the well 98 in the northwestern slope of the uplift, indicating that the rocks have experienced the shock pressure of up to 10 GPa there.

Based on observations in the core of the VDW and other boreholes penetrating into crystalline rocks, the latter have subjected to shock pressure, differential movements of rock masses, injections of impact melt, thermal metamorphism, as

Fig. 16 Shatter cones in a Neoproterozoic argillite. Borehole 98 in the west part of the central uplift

Fig. 17 Shatter cones in a biotite-amphibole gneiss. Borehole 40, depth 295 m (southern part of the central uplift)

well as hydrothermal alteration. Shatter cones are present in crystalline rocks in places (Fig. 17). The shock transformations are described in detail in the Sect. 4.1. The shock compression, which in the upper part of the VDW core is estimated to reach 45–50 GPa, decreases gradually but irregularly to 15–20 GPa towards the bottom. The evidences of the attenuation of shock pressure in radial direction are also recorded in the subsurface zone of the central uplift; the shock pressure decreases from 45–50 GPa in the center of the uplift to 30–35 GPa at 3–4 km from the center (see for detail Sect. 4.1).

The arched part of the central uplift is complicated by a central pit elongated in latitudinal direction for 5.5 km (Figs. 11 and 15). The pit is 4.2 km across, the maximum depth is 525 m. The steepness of its slopes does not exceed 20°–25°. According to drilling data, the pit is filled from bottom to top with polymict mesobreccias (up to 118–160 m thick), suevites (up to 60–102 m), coptomict gritstones and sandstones (up to 106 m), Middle Jurassic clays (up to 220 m), and Neogene to Quaternary sediments (up to 60 m). The same formations overlie the slopes of the uplift and, in the north of the latter, the uplift crest (Fig. 2). Directly upon the parautochthon, polymict mesobreccia rests everywhere. In southern and eastern and, locally, in northern slopes, it is overlain by suevites and polymict microbreccias. The summary thickness of these allogenic formations peaks 291 m (bh. 25). Both allogenic impact-derived formations and parautochthonous rocks of the uplift are covered by Lower and Middle Jurassic crater lake deposits (up to 225 m) and by Callovian to Quaternary deposits (up to 115 m). The ring crest of the uplift is exposed from under crater fill formations almost along the full length (except for a short segment at the north), but it is almost everywhere (except the highest points of no more than 0.5 km^2 of total area where Miocene sands rest directly over brecciated Neoproterozoic argillite—bhs. 48 and 158) overlain by crater lake deposits up to 105 m thick, and in the southeast, also by Middle to Upper Jurassic marine deposits (up to 65 m) and Neogene and Quaternary sediments (up to 35 m). In the west and southwest of the uplift, the deformed Neoproterozoic rocks are covered only by Neogene and Quaternary sediments.

Deep-seated zones of the central uplift were studied in detail by the VDW core; some observations were obtained from shallower boreholes. The results of these studies are presented in next sections.

4 Impact Lithic Breccias and Impactites

4.1 Crystalline and Sedimentary Authigenic Breccias

All crystalline basement and platform cover lithologies, which are penetrated by boreholes below the crater fill, are subjected to a variable degree to deformations and impact-induced transformations. They are partly displaced from their previous position and are regarded as either autochthonous or parautochthonous formations. As a result, they are attributed to authigenic monomict (crystalline or sedimentary) breccias. These breccias do occur everywhere in the zone adjacent to the surface of the crystalline basement, but details of its structure and transformation of rocks in this zone are beyond reach of studying except for the central uplift area.

The crystalline rocks are brecciated and cataclased throughout the VDW section (see Fig. 2 in Chap. 2). Brecciation and cataclasis are most intense in the upper part of the authigenic breccia massif, down to a depth of 3.5 km from its roof; below these deformations occur only at separate intervals up to 50 m long. The size of monolithic blocks commonly does not exceed a few meters, and rarely reaches several tens of meters in the upper part of the authigenic breccia, whereas down of 3.5 km from the surface of the central uplift, block size reaches several hundreds of meters n core. The intensity of brecciation and cataclasis, as well as of thermal and hydrothermal alteration of various rocks, is clearly fixed by the change of their density (see Sect. 2, Fig. 9). The gneissose structure is commonly oriented nearly parallel to the core axis, but in zones of brecciation it is disturbed; cataclasites with mylonite lenses and pseudotachylites veins are appearing there. Cataclasites are traced to a depth of 4.4 km from the roof of the uplift.

Cataclasites, mylonites and pseudotachylites are also recorded in crystalline rock fragments from allogenic breccias and inclusions in impactites. Most of these rocks are originated due to the impact transformation of crystalline rocks, but some of them may be formed much earlier during the Early Precambrian dynamothermal metamorphism. While these rocks retain shock features, their typing as impact-derived formations is evident. In addition, the decrease of cataclastic rocks abundance downward in the VDW section also indicates that the most of these rocks arose during the impact.

Cataclasites and mylonites occur together, they differing by their texture and structure. Mylonites are schistose, whereas cataclasites usually are massive. Pseudotachylites occur both as independent bodies and together with cataclasites and mylonites.

Cataclasites are dense massive breccia-like rocks consisting of deformed rock fragments and minerals ranging from a few millimeters to several tens of centimeters across. They are cemented by a finely granulated polymineral matrix with particle size up to several tenths of a millimeter. Cataclasites inherit the color of original rocks: they are of lighter tint when formed after leucocratic lithologies, but are dark-grey to black when the precursors are mesocratic or melanocratic lithologies. Cataclastic, brecciated, or ocellar structures are common. Larger fragments of initial rocks commonly are fractured and cut by veins of fine-granulated matter from the groundmass. In places, cataclasites are crushed and crumpled up to fluidal structure appearing (Fig. 18).

Mylonites occur locally within the cataclasites. By the degree of granulation, mylonite and ultramylonite are distinguished. The gradual transitions from cataclasite to mylonite or ultramylonite are characteristic (Fig. 19). Mylonites are dense, schistose rocks with lenticular stripes of rock or mineral fragments that have survived after re-crushing. They are black-colored commonly. Most of the mylonites are oriented according to gneissic banding, but in places, they can cross the latter. The predominant textures are mylonitic or blastocataclastic, and structures are schistose, banded, or lenticular. The groundmass of both mylonite and ultramylonite is composed of a fine-crushed (the particles are sized to be hundredths of a millimeter) mosaic aggregate of quartz, feldspars, mica and, in rare places, other femic minerals that form an alternation of parallel bands (strips) of several millimeters wide. These bands combine with pinches and swells composed of a coarser material.

Pseudotachylites predominate in areas of intense cataclasis, but they are also found outside of these areas, which may be due to distant displacements of certain rock blocks. The pseudotachylites form concordant and crossing veins with sharp contours, which are splitted or interrupted frequently (Fig. 20a). Their maximum thickness is 7 cm. This lithology has a glassy image, black color and, in places, striped or fluidal structure (Fig. 20b). The groundmass texture is blastoporphyritic, blastocataclastic, felsitic, or cryptic. The rock consists of fine-granulated clasts embedded in opaque glass. Relatively thick pseudotachylite veins enclose fragments of rocks and minerals of several millimeters in size, however, the fragments up to 3 cm across occur in places.

The chemical composition of pseudotachylites is very variable, it generally inherits the composition of the initial rock (Table 4), but in some cases, higher potassium, silica, Fe_2O_3/FeO ratio, and volatile components are recorded. The differences in chemical composition between the pseudotachylite and the host rock are mainly due to the presence of inclusions.

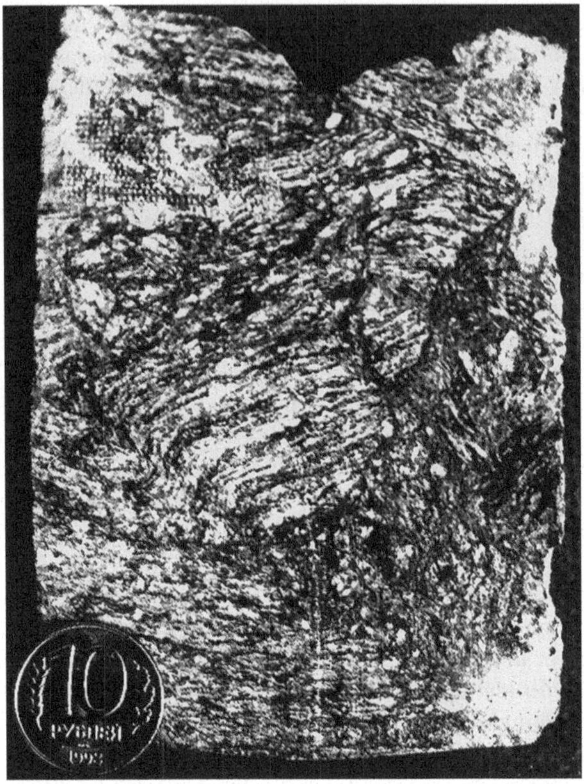

Fig. 18 Cataclasite after amphibole-biotite gneiss composed of fragments rumpled to a variable extent and cemented by fine-granular polymineral matrix (dark) (VDW, depth 920 m). Natural size

Down to a depth of 4.5 km in the VDW, the crystalline rocks are injected to a variable extent by veins of tagamites ranging from a few of centimeters to 2 m in size. The tagamite bodies are most abundant in the upper part of the authigenic breccia (to a depth of 1350 m in the VDW section) where they contribute several percent to the total rock volume within certain depth intervals (Fig. 21). Below, only rare single veinlets of no more than 3 cm thick occur. In radial direction, the common occurrence of tagamite injections in the authigenic breccia is confined within an area of the central pit, i.e. within a radius of 2.7 km.

Within this restricted area and to a depth of 3.2 km downward, the shocked basement rocks underwent subsequent thermal metamorphism. The latter is the most manifested in brecciated, cataclased and mylonitized rocks, which were experienced the shock compression from 45 to 25 GPa. The most intense thermal transformations have been recorded in the depth intervals from 890 to 1498, from 1830 to 1910, from 2400 to 2600, and from 2800 to 3050 m in the VDW. Below the degree of annealing decreases gradually with depth. The intensity of these transformations correlates directly with the degree of cataclasis and rock crushing.

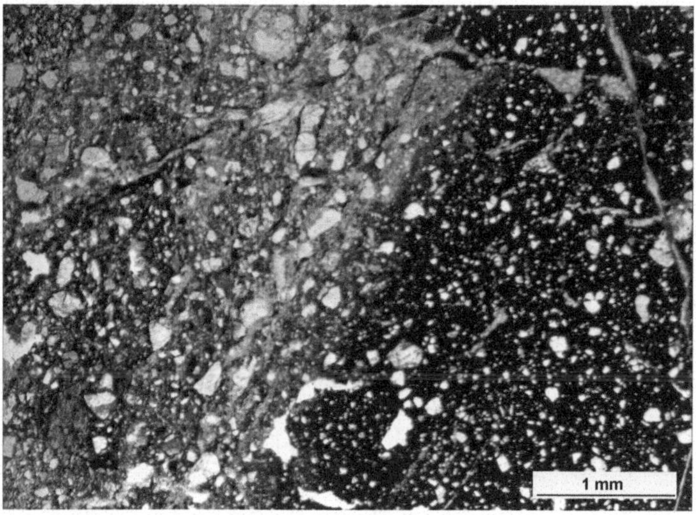

Fig. 19 Gradual transition from cataclasite (left) to mylonite-ultramylonite (right). VDW, depth 4057 m. Photomicrograph, cross-polarized light

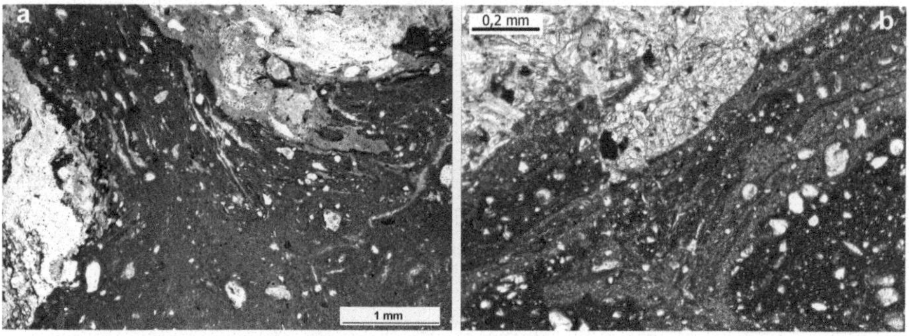

Fig. 20 Pseudotachylites cut cataclased gneisses. **a** A pseudotachylite vein with sharp contacts and lens-like accumulations of clastic material. VDW, depth 3728 m. **b** Pseudotachylite filled with small clasts, with fluidal thin-banded texture cut a. VDW, depth 3789,6 m. Photomicrographs, cross-polarized light

The authigenic breccias are nowhere penetrated by drilling in the annular trough. Based on seismic profiling data and the restored thickness of different sedimentary formations at the impact site, these breccias are assumed to be made mainly of Permian, Carboniferous and Devonian rocks. Only in western and northern sectors of the annular trough, Neoproterozoic formations compose the true crater floor. No shock features have been recorded in sedimentary (Permian and Lower Triassic) rocks composing the crater rim in different parts of the impact structure (boreholes. 31, 181, 302, I-64).

Table 4 Whole rock analyses (wt%) of pseudotachylites from the VDW core

Components	Depth of occurrence (m)									
	3833	3928	4809	3007	3125	3785-3795		4106-4157		3097
	1	2	3	4	5	6		7		8
SiO_2	61.75	52.83	53.91	58.26	55.97	54.84	50.79	48.10	46.16	51.05
TiO_2	0.61	1.46	1.11	0.76	0.94	0.96	1.09	1.21	1.06	2.12
Al_2O_3	14.81	14.81	14.91	15.38	16.11	15.62	14..55	15.55	13.78	11.27
Fe_2O_3	2.03	3.75	3.65	3.33	3.15	3.88	3.96	5.58	5.02	5.35
FeO	3.07	5.21	2.84	4.53	3.78	4.12	6.08	4.71	6.89	6.14
MnO	0.04	0.09	0.12	0.07	0.11	0.12	0.15	0.12	0.14	0.15
MgO	2.73	4.89	4.41	4.11	3.94	4.41	6.26	6.21	8.76	8.82
CaO	2.03	2.77	3.52	4.61	5.99	7.84	9.61	7.11	8.41	10.54
Na_2O	1.53	0.87	0.76	3.47	3.42	3.35	2.62	2.05	1.66	1.91
K_2O	6.99	6.71	8.62	1.11	3.02	1.74	0.99	3.65	1.10	1.30
P_2O_5	0.06	0.16	0.72	0.05	0.21	0.12	0.08	0.14	0.08	0.21
L. O. I.	4.11	6.01	4.83	4.31	3.47	2.85	3.55	4.64	6.12	0.73
Total	99.76	99.56	99.40	99.99	100.11	99.85	99.73	99.07	99.18	99.59
n	1	1	1	1	1	2	3	2	3	1

XRF data, analyzed by B. A. Tsimoshenko, VSEGEI Labs, St. Petersburg, with a SRM-2 instrument

n Number of samples analyzed

1–3, Pseudotachylites cut migmatized biotite and amphibole-biotite gneisses; 4 and 5, pseudotachylites cut biotite-amphibole plagiogneisses; 6 and 7 pseudotachylites in dolerites; 8, pseudotachylite in amphibolites

Fig. 21 Brecciated and cataclased shocked amphibole-biotite gneiss with thin tagamite veins. VDW, depth 1250.5–1255.5 m. Core diameter is 73 mm

4.2 Sedimentary Allogenic Breccias

Sedimentary allogenic megabreccias and mesobreccias cover the authigenic breccias over the entire area of the impact structure except for the crest of the central uplift (Fig. 1). A general characteristic of this lithological unit was given in Table 1.

Sedimentary megabreccias fill the annular trough, where they have been penetrated by 20 wells. Their maximum drilled thickness is 751 m (well 191), but the true thickness (see Sect. 1), can reach 1500 m in the western part of the annular through (Fig. 22). Downward these megabreccias are passing gradually into weakly the disturbed bedrock formations represented, more probably, by Neoproterozoic and Devonian rocks, and, locally in the southern sector, by Carboniferous and Permian rocks. The megabreccias are overlain by polymict allogenic mesobreccia in the central part of the western sector of the annular trough, whereas in eastern and northern sectors, only by clays of the Kovernino Formation. Closer to the central uplift, at a distance of 3–5 km from the crater center, the megabreccias rest on steep slopes of the uplift.

The megabreccias consist of deformed blocks of sedimentary rocks (from Neoproterozoic to Permian) up to several hundreds of meters in size and cemented by fine-crushed material of the same rocks. In many cases (boreholes 16, 145, 191, 200), an inverse stratigraphic sequence has been recorded due to random alternation of blocks of different formations. For example, in the well 191 Carboniferous dolomites are overlain by a 419 m thick Upper Devonian sequence, a thick zone of brecciated rocks dividing these two formations (Bogorodskaya and Tumanov 1980).

Sedimentary variegated mesobreccia. This block-rubble breccia consists of fragments of Lopingian Permian and Lower Triassic rocks. It occurs everywhere in the ring terrace and locally, in near-slope parts of the annular trough. It is penetrated

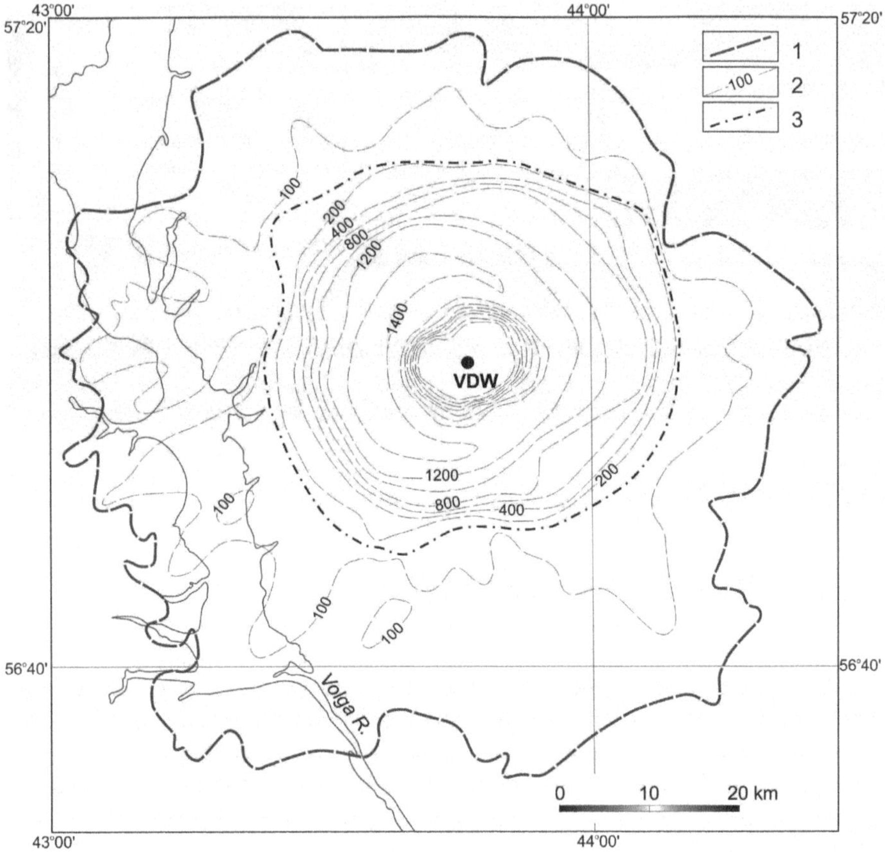

Fig. 22 Isopach map for sedimentary megabreccia and variegated mesobreccia. 1, outer limit of sedimentary megabreccia and variegated mesobreccia extension; 2, contour lines of summary thickness of sedimentary megabreccias and variegated mesobreccia; 3, outer boundary of the annular trough. Contour interval is 100 m at the ring terrace, and 200 m in the crater depression. VDW is shown by a double circle

by more than 170 wells and exposed in the western part of the impact structure. Within the outer ring terrace, its average thickness is about 100 m, but in axial parts of radial trenches, it reaches 200 m (well I-22). Inward to the crater, it increases peaking 281 m (bh. 31) in peripheral parts of the annular trough adjacent to the ring terrace (Tumanov and Bogorodskaya 1975). The variegated mesobreccias rest on dislocated Permian rocks, and in the northern and western sectors of the impact structure, on the Lower Triassic (Blom 1960; Tumanov and Bogorodskaya 1975).

The variegated mesobreccia has been studied in detail in the western sector of ring terrace, where their numerous exposures are traced for more than 50 km along the western bank of the Gorky Reservoir (Fig. 13). The main body (90 vol.%) of this mesobreccia consists of blocks and fragments of local (Lopingian and Induan) variegated clays and siltstones, less often marls and sandstones; the size of fragments is up to a few meters across. The rest (10%) is a material originated from the deeper stratigraphic levels of bedrock, mostly from Upper Carboniferous and Cisuralian carbonate lithologies. Both the amount and size of these rock fragments decrease in radial direction. As it shown by borehole sections located close to the junction of ring terrace with annular trough edge, carbonate rock fragments contribute occasionally more than 60 vol.% to the variegated mesobreccia in this zone (Bogorodskaya and Tumanov 1980). An accumulation of large blocks (of up to 50 m across) of Pennsylvanian limestones in a quarry near the village of Bol. Babie occupies an area of about 2 km^2. This cluster of carbonate blocks as well some other accumulations in the inner zone of the ring terrace may be regarded as the remnants of ballistically ejected material. Along the outer edge of the ring terrace, a higher amount of carbonate rock fragments (up to 10 vol.%) occur only within the radial troughs, marking areas where the farthest transportation of ejected material took place (Fig. 23). In rare cases, mortar structure and shatter cones are observed in carbonate fragments there (Fig. 24).

Lumps of Lopingian and Induan clays are randomly distributed in a homogenized psammitic to aleuritic groundmass containing small (up to 3 cm across) fragments of sandstones, marls etc. Most of fragments display traces of dragging, crushing, and rumpling. Some rounded denser fragments are mantled by loose arenaceous material (armored blocks); such fragments are particularly common north of the

Fig. 23 A large block of Carboniferous limestone in variegated mesobreccia composed of crushed and crumbled fragments Permian red clays. Western bank of Gorky reservoir

Fig. 24 A shatter cone in a limestone fragment from the variegated mesobreccia. West bank of Gorky Reservoir, near Kulaevo village

town of Puchezh where the axis of the most strongly pronounced radial trough is located (Fig. 25).

The heavy mineral fraction extracted from the cement matter of the variegated mesobreccia includes epidote (42–73%), magnetite and ilmenite (10–40%), garnet (3–13%), etc. This mineral assemblage is common in Lopingian and Induan formations (Bogorodskaya and Tumanov 1980). However, in samples from the axial part of the Puchezh radial trough, pyroxene and hornblende, which are not characteristic for these rocks, contribute up to 10%.

Downward in the section, the increase of size of blocks constituting the variegated mesobreccias occurs. Klippen of Lopingian rocks in the western sector of the ring terrace range from several tens of meters to a few kilometers across. In radial direction, they are replaced by a zone of intensely dislocated rocks with numerous upthrusts, overthrusts, and various folds. This zone is the most pronounced in the bluffs south of the town of Chkalovsk.

In order to reveal distribution regularities of the displaced material composing the variegated mesobreccia, a statistical processing in axonometric charts of bedding orientation in platy lumps and blocks has been carried out in exposures along the banks of the Gorky Reservoir. These blocks range from 1 to 100 m across (for larger blocks, several measurements of orientation were taken). The tabular lumps and blocks are usually oriented conformably to their bedding and are suggested to reflect a direction of their displacement (Fig. 26). The results of the statistical

Fig. 25 Variegated mesobreccia comprising rounded fragments of dark-red claystones coated by grey-colored loose arenaceous material. Western bank of Gorky reservoir

Fig. 26 Variegated mesobreccia consisting of block- and rubble-sized fragments of Permian and Triassic claystones, siltstones, and sandstones. Flattened blocks of dense rocks dip left and indicate the laminar movement of a mudflow. Western bank of Gorky reservoir

processing are presented on the stereograms for layering elements in blocks (or—what is the same—orientations of plate-shaped blocks) measured for different areas in the western sector of the ring terrace (Fig. 27). In its central part, where the axis of the Puchezh radial trough is located, and thus the upper part of the mesobreccia unit is outcropped, the maximum frequency of the predominant orientations is only 14%, and the amount of steeply dipping (angles of inclination more than 45°) blocks exceeds 34%. In the northern and southern sectors, where the lower, more coarse-fragmented mesobreccia is exposed, the maximum frequency of orientations reaches 37%, whereas the amount of steeply dipping blocks is as low as 12%. Downwards, the orientation of fragments is more homogeneous, indicating a gradual change from turbulent movement to a laminar regime during formation of this rock sequence.

The second result of the structural analysis is the predominance of bedding strike azimuths close to radial directions. Loadings on the predominant directions vary in the distinguished areas similar to the frequencies of orientations. As a rule, elongated

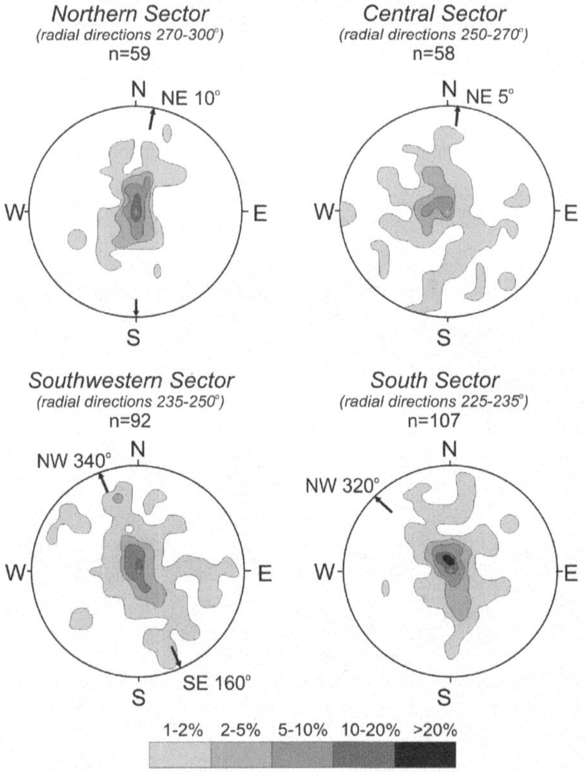

Fig. 27 Stereograms demonstrating orientation of plate-shaped fragments of clayey rocks from variegated mesobreccia in the western sector of the ring terrace. Results of measurements of linear features (bedding, orientation of contacts) are given for different parts of the western sector. At stereograms, frequencies of dipping orientations are shown by contour lines, and predominant directions of dip azimuths, by arrows

clay lumps of 0.5–1.0 m across embedded in a fine-grained clastic cement show a radial orientation relative to the crater center (see Fig. 13).

The sequence of the variegated mesobreccias formed mainly by radial transport of water-saturated debris that had a character of mud-rock flows, particularly in the final stage of movement. The movement was laminar near the bottom, and turbulent in the upper part of the flows. Some material may have been thrown out along ballistic trajectories. The presence of blocks of Upper Carboniferous limestone in the ring terrace west of the town of Puchezh indicates such a mode of transportation over a distance of up to 22 km.

The sedimentary megabreccias and mesobreccias described above are overlain by a blanket of ballistic ejecta and sediments that settled from the explosion cloud. This blanket up to 291 m thick includes polymict mesobreccias, suevites and polymict microbreccias. Within the blanket, the above lithologies are linked by gradual transitions. The most of the ejected material occurs in the northwestern sector of the impact structure (Table 2). In general, the centers of gravity of the preserved polymict mesobreccias and suevites are displaced by about 3–4 km to the northwest (310°–340°).

4.3 Polymict Allogenic Breccia

Polymict mesobreccia consisting of fragments of crystalline and sedimentary (Neoproterozoic and Devonian) rocks is penetrated by 57 boreholes. In the central uplift it rests on crystalline authigenic breccias, and in the annular trough, on sedimentary megabreccias. In the upper part of the section, they pass gradually into suevites in the central part of the crater or into polymict microbreccia in the annular trough. The lower boundary of polymict mesobreccia ranges from −435 m (bh. 38) to +11 m (bh. 746).

The thickness of this breccia is mainly controlled by the relief of the underground; it reaches 170 m (bh. 25) at the southern slope of the central uplift. In the central pit, the thickness of this formation peaks 125 m in the VDW. It decreases sharply towards the dome part of the uplift; at the crest of the uplift, it is absent (Fig. 28). In the annular trough, the polymict mesobreccia commonly is of 10–30 m thick, but in its southern part, it reaches 70 m thick (bh. 70). Nevertheless, the bulk of the preserved volume of polymict mesobreccia (ca. 80%) occurs in the annular trough owing to its large distribution area.

The clastic material in polymict mesobreccia is represented by both crystalline and sedimentary rocks. Among the fragments of crystalline rocks, both unshocked and variably shocked (up to stage II) amphibole-biotite gneisses and amphibolites as well as cataclasites and protoimpactites dominate. Fragments of sedimentary rocks are represented by various limestones (cryptocrystalline, marbled, organogenic etc.), dolomites, marls (often with “gries” structure), clays, shale, sandstones, argillites and siltstones. The clastic material often contains minor (<1%) admixture of small fragments of impact glasses. The latter occur also in the matrix, which is represented by a fine-grained material of silicic, clayey, or carbonate-clayey composition.

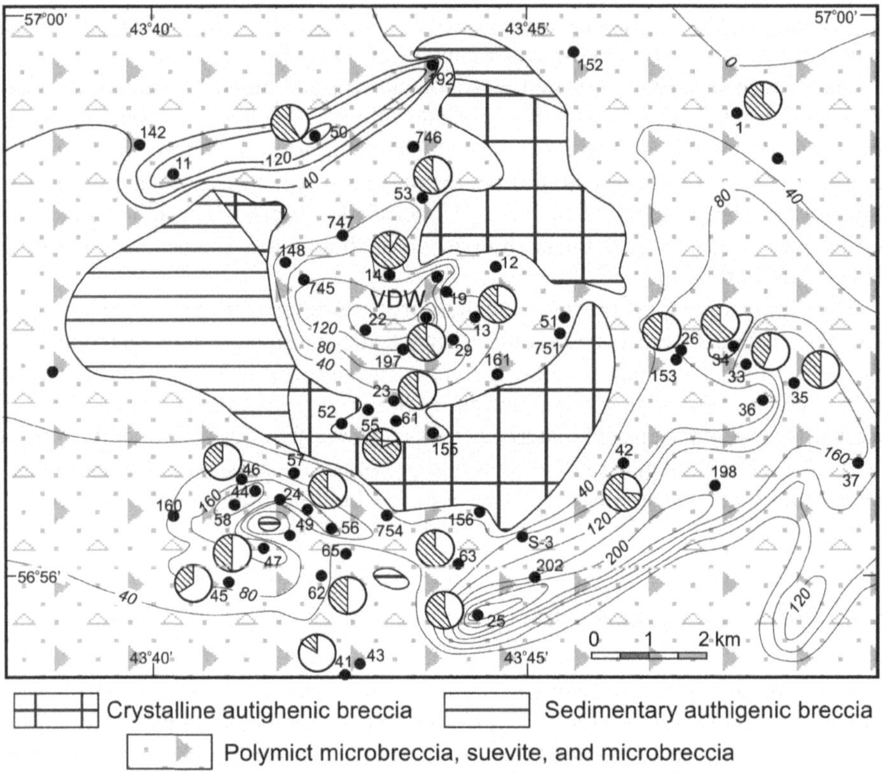

Fig. 28 Isopach map for ballistic ejecta formations (polymict mesobreccia, suevites, and microbreccia) in the central part of the Puchezh-Katunki crater. Isopachs show the summary thickness of three above formations based on borehole logs, core study, and gravimetric data. Contour interval, 40 m. Boreholes penetrated polymict breccias and suevites are shown by black numbered circles. For some boreholes, a circle diagram showing a fraction of crystalline rock fragments (*shaded*) against clasts of sedimentary rocks is given. The VDW is shown by a double circle

The abundance of crystalline and sedimentary rock clasts varies over a very wide range. This indicates to the irregular distribution of ejected material (Fig. 28); however, a general trend is outlined, it consists in the decreasing of crystalline rock fragments in centrifugal direction.

Polymict breccia commonly is a dark grey lithology, which is characterized by an irregular distribution of fragments with respect to abundance, composition and size (Fig. 29). A typical section of this formation is described in the depth interval from 432 to 550 m in the VDW. A rough grading of clastic material is outlined: the average size of fragments is increasing with depth from 3 to 5 cm (with variations from 0.3–0.8 to 5–20 cm) to 20–50 cm below 520 m. In addition, the relative amount of both fragments of sedimentary rocks and impact glasses decreases downward in the section. The sedimentary fragments are black and brown argillites, siliceous-clayey shales, and less often siltstones. Fragments of crystalline rocks are usually

Fig. 29 Polymict allogenic mesobreccia comprising approximately equal fractions of crystalline and sedimentary rock fragments up to 3 cm across. Shocked or, more rarely, non-altered gneisses (white) and dark-colored claystones are predominant among clasts. Borehole 65, depth 170 m. Core diameter is 81 mm

shock-metamorphosed up to stage II and have a mortar structure. Larger fragments are cut in places by veins of pseudotachylites. In some intervals, i.e. between 478 and 496 m, and between 517 and 526 m, small particles of impact glass contribute 1–5 vol.% to the breccia. They are occasionally recorded in the mesobreccia down to a depth of 544 m. Both the clasts and the matrix are affected by alteration, which is marked by the appearance of smectites, hydromica, rare zeolite-calcite veinlets and fine pyrite dissemination.

On the southwestern slope of the central uplift (in the vicinity of the village of Emokhony), the polymict mesobreccias are of somewhat another appearance. There, around the protrusion of the Neoproterozoic rocks of about 220 m high, in immediate vicinity of the contact (most likely, tectonic) between Neoproterozoic argillite and Neoarchean crystalline rocks, a breccia consisting of intensely deformed blocks (from 1 to 100 m across) of both above-mentioned formations occurs. The rocks composing the blocks display shatter cones and mortar structure. This breccia is cemented by vitro-lithoclastic suevites or, less often, by tagamites and fine-clastic breccias. This cement makes up to 40 vol.% of the breccia. Similar accumulations of large blocks at the base of the polymict mesobreccia seem to occur in other parts of the central uplift too, in particular in the eastern part of the central pit where blocks of sedimentary rocks up to 34 m across rest on parautochthonous Neoarchean-Paleoproterozoic crystalline rocks (e.g., in boreholes 17 and 29).

Polymict microbreccia crowns the section of the crater fill. It overlies suevites at the central uplift, and polymict mesobreccias in the annular trough, some microbreccia lenses lie within suevite or polymict mesobreccia, though. The microbreccias opened by 15 boreholes. Within the central uplift, it develops in separate spots, whose distribution shows no clear correlation with the relief of its base indicating the initial inhomogeneity of their distribution. Their average thickness in the central uplift area is about 10 m with maximum values 34 m (bh. 42) and 29 m (bh. 43). The bulk of this formation (95%) occurs in the annular trough where its thickness reaches 80 m (bh. 205).

The polymict microbreccia is a low-density non-layered lithology. It consists of fragments of sedimentary and crystalline rocks ranging commonly from 0.2 to 0.5 cm across (Fig. 30), some fragments (2–3 vol.%) are up to 7 cm across, though. An admixture of about 1 vol.% of small impact glass particles can occur in places. Both size and amount of clasts increase downwards in sections. Proportions of different lithologies among the clastic material vary considerably as in the polymict

Fig. 30 Polymict microbreccia consisting of rock and mineral fragments of <2 mm across cemented by clayey fine-crushed material. The cement is somewhat altered to clay minerals causing a greenish tint of the rock. Borehole I-64, depth 150 m. Core diameter is 81 mm

mesobreccias. For example, in borehole 42, crystalline rocks contribute 80% of the clasts, but in the borehole 43, only 15%.

To characterize this formation, the distribution of clastic material by granulometry and petrographic composition for a microbreccia occurring on the southeastern slope of the central uplift (borehole 42, depth of 210–240 m) is given (Table 5). In addition to data presented in this table, the heavy mineral fraction in granulometric classes of $-1+0.5$ and -0.5 mm includes amphibole (74.6 and 77.7%, respectively), pyroxene (10.1 and 8.5%), biotite (8.0 and 6.6%), garnets (1.4 and 4.2%), sulfides (0.7 and 0.8%), magnetite (0.7 and 2.2%), epidote (2.2%, only in the class $-1+0.5$ mm). The light fraction comprises chlorite, feldspars, and quartz together with some calcite (up to 0.5 vol.%).

The crystalline clasts are both unshocked and shock-metamorphosed; some larger fragments bear shatter cones. The matrix consists of pelitic material. It is of clayey, clay-siliceous, in places clay-carbonate composition in dependence on predominance of certain sedimentary in the clastic constituent. If argillites or claystones prevail among the latter, the matrix contributes up to 60 vol.%.

4.4 Impactites

Impactites, i.e. impact-derived lithologies consisting completely or partially of impact melting products contribute in the Puchezh-Katunki much less than in many other large terrestrial craters, e.g. Popigai (Masaitis et al. 1998; Masaitis 2019) or Kara (Selivanovskaya et al. 1990). They were shortly characterized elsewhere (Masaitis et al. 1980, 1995; Feldman et al. 1984; Selivanoskaya 1992 etc.).

The term "impactite" is applied to the group of impact rocks composed completely or to a significant extent (>10 vol.%) of chilled or crystallized silicate impact melt, with (or without) admixture of lithic and mineral clasts (Masaitis et al. 1980, 2018; Masaitis 2005, etc.). This restricted use of the term is in keeping with its original meaning (Murawsky 1977; Bates and Jackson 1987; Chao and Xie 1990). This group includes two rock species—tagamite and suevite. Tagamite is composed mainly of chilled or crystallized impact melt with admixture of mineral and lithic clasts, while suevite consists of lithic, mineral and glass fragments. Further subdivision is based on such petrographic and lithological criteria as fabric, crystallinity of groundmass, ratio of lithic and mineral clasts with respect to glass particles, grain size, shape and composition of fragments, type of cementation, etc. A term "protoimpactite" is proposed for strongly shocked crystalline rocks that preserve their initial structure, but are composed mainly of chilled or recrystallized monomineralic shock melts, diaplectic glasses and relics of pristine minerals.

Suevites are rather widespread in the Puchezh-Katunki structure. They are closely associated with polymict allogenic breccias. Generally, suevites overlie the polymict mesobreccia and, in places, are covered by either polymict microbreccia or lenses of rewashed material—coptomict gritstones and sandstones. The transition from suevites to coptomict gritstones is usually gradual and established by the appearance

Table 5 Distribution of rock, mineral and glass particles in microbreccia by granulometry and by composition (Borehole 42, depth interval 210–240 m)

Components	Granulometric classes (mm)							
	+10	−10 + 7	−7 + 5	−5 + 3	−3 + 2	−2 + 1	−1 + 0.5	−0.5
Amphibole gneisses	62.8	63.1	65.7	63.1	57.1	61.5	33.8	15.4
Amphibolites	29.0	19.8	16.1	15.8	14.5	7.7	3.1	–
Granite gneisses	–	6.8	5.5	6.4	7.1	5.7	–	–
Siltstones	1.2	6.8	7.2	11.8	16.4	21.8	23.3	14.8
Sandstones	–	–	2.7	–	–	–	–	–
Limestones	–	–	2.8	1.3	1.9	–	1.1	1.3
Impact glass	7.0	2.3	–	0.4	–	–	–	–
Protoimpactite	–	1.2	–	1.2	3.0	3.3	4.9	1.7
Plagioclase + quartz (7:3)	–	–	–	–	–	–	33.8	66.8
Yield of material separated to granulometric classes (wt%)	43.80	27.80	22.10	5.00	1.00	0.20	0.050	0.006

Initial weight of the analyzed sample after quartering is 1.17 kg of visible signs of rewashing, as well as by a sharp increase of the amount of matrix and change in the composition of cement matter. The lower boundary of the suevite sequence can be traced quite clearly by a sharp decrease of the amount of impact glass fragments accompanied by increase of the size of individual crystalline rock fragments.

The suevites are opened by 35 boreholes within the central uplift area, and by 4 boreholes in the northwestern outer slope of the annular trough at 20–23 km from the crater center. They occur as lenses in the lows of surface relief of the polymict mesobreccia or as small irregular bodies inside the latter; in places, suevites compose a cement matter of this breccia. The drilled thickness of suevites varies from a few meters to 121 m (bh. 25).

The suevites are generally dense, greenish-grey rocks consisting of clasts of crystalline (amphibolites, gneisses, shists, etc.) and sedimentary (argillites, siltstones, sandstones, limestones) rocks with variable ratios between them, as well as of impact glass fragments and of clasts of minerals (Fig. 31). In dependence on ratios of different clastic constituents, litho-vitroclastic, vitro-lithoclastic and vitro-crystallo-lithoclastic suevites may be distinguished. The average size of lithoclasts is 0.5–1.5 cm, of vitroclasts, from 1.5 to 3.5 cm, and of crystalloclasts, 1–2 mm and less. Individual gneiss fragments reach as much as 10 cm, and vitroclasts, 5 cm. The vitroclasts are rounded, elongated or curved, frequently with irregular undulated outlines or with ragged edges. Among impact glass fragments, light yellowish or greenish porous to slag-shaped glasses with inclusions of lechatelierite, shock-metamorphosed quartz and feldspars are predominant. Massive greenish-brown glasses with lump structure also occur. The most of glasses are almost completely

Fig. 31 Lapilli litho-vitroclastic suevite with rare lithic clasts. VDW core, depth about 380 m

altered. According to thermo-gravimetric and X-ray data the altered glasses consist of montmorillonite (50–60 vol.%), and calcite (15–20 vol.%) with small admixtures of feldspar, quartz, and amphibole clasts Among crystalloclasts, rare graphite flakes and impact diamonds are found in suevites.

In general, the local fields of suevites distribution in the impact structure differ from one another. Suevites most enriched in impact glass occur within the central pit area and on the outer slope of the annular trough.

The suevite area in the central pit is 3.0 km long and 1.9 km wide (Fig. 2). There, suevites are penetrated by 13 boreholes including the VDW; they form a lens with maximum thickness of 101 m (bh. 22). All suevite varieties occur in this distribution field. The litho-vitroclastic suevites (Fig. 32) are best studied in the VDW section. The average content of vitroclasts in this lithology is 34.2 vol.%, with variations from 21.3 to 44.8 vol.% The amount of sedimentary and crystalline rock fragments is about 20 vol.%, and of crystalloclasts, about 3 vol.%, The average matrix content ranges from 13.6 to 25.4 vol.% (20.3 vol.% on the average). The vitro-lithoclastic suevites are characterized by equal amounts of crystalline and sedimentary rock fragments (Fig. 33); in places, clasts of protoimpactites and mylonites are present. Angular fragments of impact glasses reach 15 cm across. The summary amount of vitroclasts does not exceed 10–15 vol.%. Vitro-crystallo-lithoclastic suevites are rare; they are drilled only by boreholes 29 and VDW. They are characterized by the prevalence of crystalline rocks amongst the lithoclasts, and high contribution (up to 30 vol.%) of crystalloclasts (both unshocked and shock-metamorphosed clasts of amphibole, biotite, pyroxenes, feldspars and quartz, as well as of their intergrowths); vitroclasts amount for 10% (Fig. 34).

The suevites are cemented by clayey, argillo-siliceous matter, with abundant fine particles of modified glass and quartz fragments. In litho-vitroclastic suevites, hydrothermal cement composed of zeolites and calcite occurring as overgrowths of individual crystals is observed in places. In addition, zeolites and calcite fill veinlets and pores in glasses (Naumov 2002).

The average chemical compositions of suevites from VDW and from shallow boreholes in the central pit are similar (Table 6), but differ in calcium and sodium contents owing to the different amount of fragments of carbonate rocks. However, the high values of dispersion for the most of major chemical components indicate the variability of composition of suevites from certain occurrences.

At slopes of the central uplift, suevites extend toward south and east where they distribute within an arched band of about 12 km long and 1.5–2.5 km wide. This distribution field extends eastward to the annular trough where the thickness of suevites does not exceed 53 m (bh. 38). In addition, a small isolated occurrence of suevite (1.0 × 0.5 km in size) is situated in the northwest of the central uplift (Fig. 2). The thickness of suevites at uplift slopes varies sharply; this is probably due to accumulation of ejecta fractions enriched by impact melt fragments within structural lows caused by terrace-like subsidence of the slope. In these lows, rather thick, elongated along the uplift slope suevite lenses (up to 72 m thick) are divided by areas where the suevites are not more than 15 m thick.

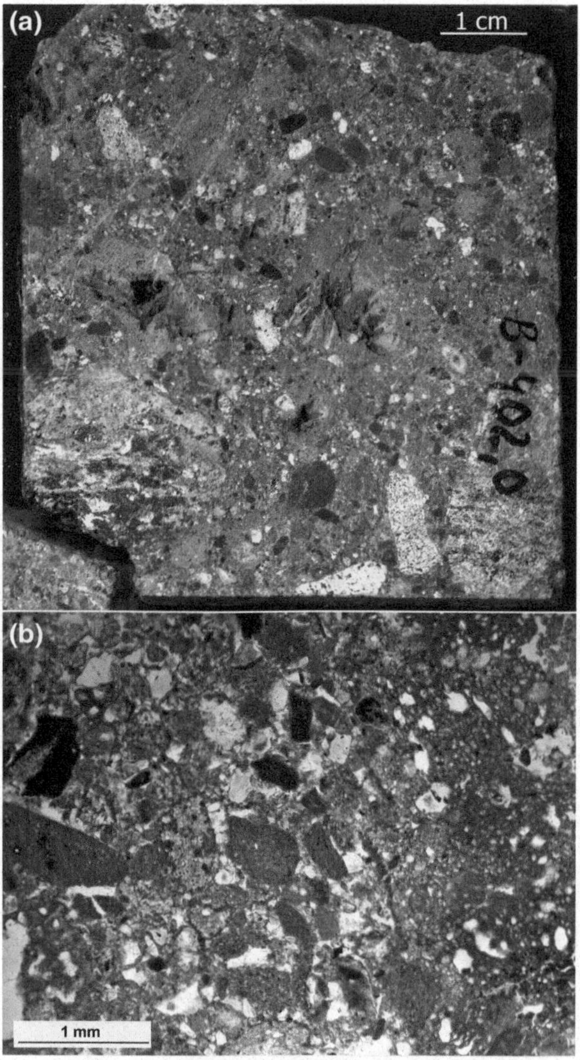

Fig. 32 Litho-vitroclastic suevite from the central pit. **a** Polished core sample. **b** Photomicrograph, plane-polarized light. **a** Greenish (due to alteration) xenomorphic or, more rarely, automorphic impact glasses contribute about 30%. Fragments of shocked crystalline rocks are much more abundant and larger than of sedimentary ones (mostly dark argillites). The latter are no more than 0.6 cm across, while crystalline clasts and impact glasses reach 3 cm. **b** A large fragment of a porous greenish impact glass filled with crystalline rock and mineral micro-inclusions in litho-vitroclastic suevite (at right). The groundmass is composed of angular clasts of claystone, argillite, marl, quartz, feldspar, crystalline rocks. VDW, depth 402 m. Core diameter is 65 mm

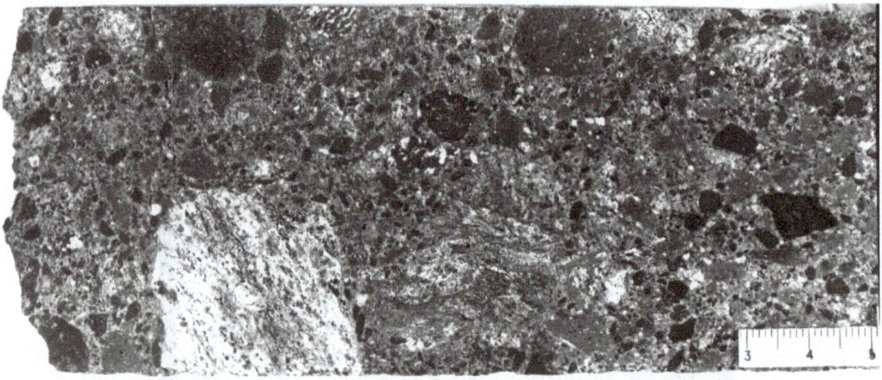

Fig. 33 Vitro-lithoclastic suevite. Although larger (up to 3 cm across) clasts are represented by impact glass or shock-metamorphosed gneisses, the total of sedimentary clasts amount crystalline rocks. Polished core sample. VDW, depth 390 m. Core diameter is 65 mm

Fig. 34 Vitro-crystallo-lithoclastic suevite. Larger fragments of gneisses, siltstones, and slag-like glasses (Gs) distribute evenly within the groundmass consisting mainly of quartz and feldspar crystalloclasts. Impact glasses contribute no more than 10 vol.%, they are greenish or brown due to clayey alteration. VDW, depth 419 m; cross-polarized (left) and plane-polarized (right) light

Table 6 Average composition (wt%) of suevites and coptomict gritstones from the Puchezh-Katunki impact structure

Components	1		2		3	4	5	
	X	S	X	S	X	X	X	S
SiO_2	59.20	0.87	58.18	3.52	56.41	54.26	58.83	1.13
TiO_2	0.70	0.04	0.55	0.21	0.71	0.75	0.70	0.05
Al_2O_3	15.42	0.53	14.53	0.97	14.20	13.03	15.62	1.20
Fe_2O_3	3.94	0.35	3.65	1.39	2.13	4.75	4.51	0.76
FeO	1.80	0.52	2.47	0.46	4.51	1.26	0.74	0.75
MnO	0.13	0.02	0.13	0.04	0.18	0.05	0.05	0.05
MgO	3.31	0.25	3.60	0.85	3.55	4.07	2.38	0.73
CaO	3.68	0.49	5.62	3.29	4.24	6.08	3.73	0.55
Na_2O	4.79	0.68	2.44	0.62	2.71	0.49	3.77	0.39
K_2O	1.88	0.15	1.24	0.36	1.22	1.57	1.52	0.19
P_2O_5	0.09	0.03	0.19	0.08	0.17	0.11	0.13	0.03
Cr_2O_3	0.01	–	0.04	–	0.14	0.01	–	–
L.O.I.	5.03	1.17	7.60	2.42	9.45	14.03	8.05	1.73
Total	99.98	–	100.24	–	99.62	100.46	100.03	0.25
n	9		8		4	2	9	

X mean value; *S* standard deviation
XRF data, analyzed by B. A. Tsimoshenko, VSEGEI Labs, St. Petersburg, with a SRM-2 instrument.
n is number of samples analyzed
1 and 2—suevites from the central pit (1, from VDW; 2, from boreholes 14, 19, 161); 3—suevites from the eastern slope of the central uplift (borehole 33, data presented by O.S. Bogatyrev); 4—suevites on the outer slope of the ring trough (borehole I-64); 5—coptomict gritstones from the central pit (VDW core)

Suevites at slopes of the uplift belong to vitro-crystallo-lithoclastic and vitro-litho-lithoclastic varietes. The former is more widespread; it consists mainly of fragments of amphibole-biotite and amphibole gneisses and of their minerals. Fragments of both impact glasses and sedimentary rocks do not exceed 10–20 vol.%. Similar suevites occur also in the borehole 38, located in the annular trough. The latter variety is distinct by a higher amount of fragments of sedimentary rocks (mostly clayey limestones) relative to crystalline ones. Biotite-amphibole gneiss fragments occupy no more than 10–15 vol.% and only 10% of fragments are vitroclasts. In suevites from the northwestern sector of the outer slope, fragments of crystalline rocks are variable in composition: along with amphibolites and amphibole-biotite gneisses, leucocratic biotite gneisses, pegmatites and granite gneisses are found there. Vitroclasts amount up to 30 vol.%.

Suevites from the eastern slope of the central uplift do not show any significant distinction in chemical composition from the similar rocks from the central pit. However, a slightly lower silica content and high amount of volatiles indicate indirectly an enrichment by carbonate rock fragments (Table 6).

In the outer northwestern slope of the annular trough, the suevites occupy an irregularly shaped area of about 50 km^2 (Fig. 1). Their drilled thickness ranges from 13 to 40 m, suevites belong to a litho-vitroclastic variety with a high amount (up to 60 vol.%) of altered impact glasses. Inferred from some features of its chemical composition, namely the highest values of calcium and volatile contents (Table 6), these suevites contain a maximum for this lithology amount of carbonate rock fragments.

The occurrence of suevite in the outer slope of the annular trough, as well as the considerable volume of redeposited suevites (coptomict gritstone) indicate that inside the crater, a portion of the suevite cover was probably removed in Pre-Kovernino time to give a material for coptomict deposits. At the same time, the lack of suevites in some topographic lows as well as discrepancies between isopachs and contour lines of lower boundary for the suevite sequence point to a scattered original distribution of suevites, some block displacements on the central uplift slopes that has occurred prior to the deposition of crater lake sediments could contributing as well, though.

The average content of impact glass in the suevites is assumed to be 25–30 vol.% in the central pit, 14% at slopes of the central uplift, and 30% at the outer slope of the annular trough. Based on estimated volumes of suevite in different sectors of the crater (Table 4), the preserved volumes of dispersed impact melt in the suevites are calculated to be 0.06, 0.15, and 0.37 km^3 respectively, yielding a total volume of 0.58 km^3.

Tagamites are penetrated by 41 boreholes in the central uplift area. They occur locally in two different positions. Firstly, they form lenticular bodies and thin veins in the authigenic breccia (Fig. 35). Secondly, irregularly-shaped tagamite bodies occur in the allogenic polymict mesobreccia in the southwestern slope of the central uplift.

The distribution of tagamites in the authigenic breccia is irregular as shown by sections of different boreholes (VDW, 745, S-3, 13, 14 and some others). Tagamite bodies are recorded within a radius of 2.5 km from the crater center. In the VDW section, the tagamites were recorded in the depth range from 560 to 4810 m They are more abundant in some depth intervals. Between 609.6 and 775.1 m depth, tagamites occupy 4.7 m of core, with a thickness of individual bodies from 0.1 to 3.0 m; between 907.6 and 1220 m, altogether tagamites take 9.9 m, individual bodies are from 0.04 to 2.1 m thick. Fine veinlets (from a few centimeters to less then a centimeter thick) were recorded in the depth range from 3006 to 4809 m. The total thickness of tagamites in the 4.5 km VDW section is about 21 m (0.5 vol.%). However, an estimation tagamite distribution in various areas of the central uplift showed that they occupy ca. 1.1 vol.% of drilled authigenic breccia intervals in the central pit area and ca. 0.73 vol.% in slopes of the uplift. Within the allogenic polymict breccia at the southeast slope of the uplift tagamite occupy ca. 1.6 vol.% of drilled sections, but this area is very restricted; thus, the volume of impact melt there does not exceed 0.005 km^3. Therefore, the most of impact melt is comprised within the authigenic breccia of the central uplift. Inferred from tagamite distribution in the upper part of the uplift, the total volume of preserved tagamites is estimated to be from 1.4 to 2.4 km^3.

Taking into account the amount of impact glasses fragments within suevites and coptomict gritstones, the content of melted material in all formations (tagamites,

Fig. 35 Thin veins (no more than 3 cm thick) and spots of tagamite in shocked gneiss. VDW core, depth 1250–1251 m. Core diameter is 73 mm

suevites, polymict allogenic breccias and coptomict gritstones) and preserved volumes of all listed lithologies in the impact structure, the volume of impact melting products in the Puchezh-Katunki crater is estimated to be of 2.5–3.5 km^3.

The tagamites are dark grey to black lithologies. Their contacts with host shock-metamorphosed crystalline rocks, which are often brecciated and cataclased, are sharp (Fig. 36). The tagamites enclosed in the authigenic breccia show holohyaline, cryptocrystalline, hyalopilitic and hemicrystalline textures.

Tagamite with a holohyaline texture forms thin branching veins in authigenic breccia. This is an only variety found at depths below 3 km. Their holohyaline groundmass is black or brownish-green depending on the degree of alteration (Fig. 37). The structure is lumpy, lenticular-banded, heterotaxitic, and in places fluidal. Inclusions amount from 5 to 25 vol.%; they are represented by fragments of transformed minerals and their aggregates (granulated pyroxene, mosaic quartz, fused plagioclase, opacitized biotite, and recrystallized diaplectic feldspars). Outlines of these clasts are sharp, no reactionary rims are observed. Rare rounded pores are filled with saponite, which also develops in the groundmass together with calcite and zeolites. In some places, pyrite impregnations occur.

Tagamites with cryptocrystalline, hyalopiltic, and hemicrystalline textures form relatively large irregular bodies and veins ranging from 0.5 to 3 m thick (bhs. 28,

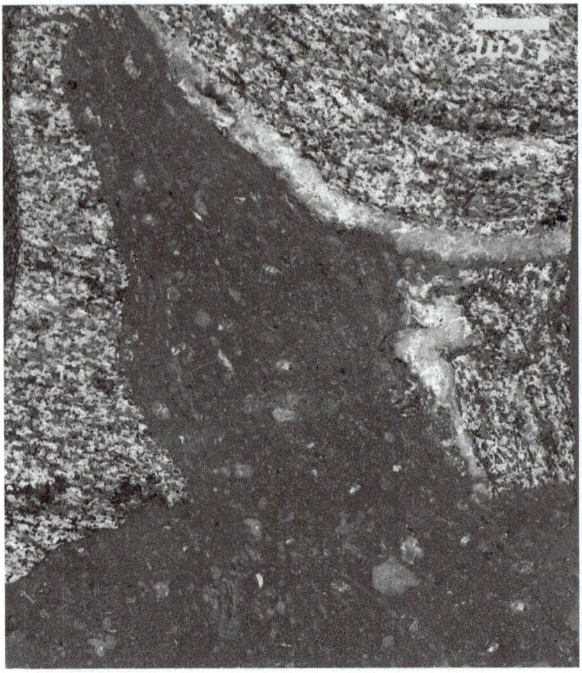

Fig. 36 Tagamite veinlet in the authigenic breccia (augmented left core from Fig. 35). The tagamite has sharp contact and contains numerous small fragments of shock-metamorphosed gneisses (VDW, depth 1251 m; full size)

Fig. 37 Tagamite with holohyaline texture and fluidal structure of groundmass comprising fragments of shocked rocks and minerals. VDW, depth 776 m

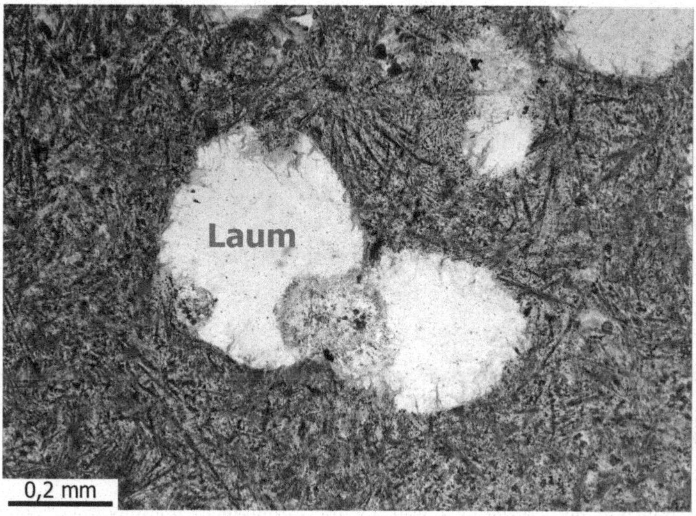

Fig. 38 Tagamite with amygdaloidal structure and groundmass stellate texture. The vugs are filled with laumontite. VDW, depth 2410 m. Photomicrograph, plane-polarized light

748, 754, VDW, etc.). Cryptocrystalline or hyalopiltic texture is recorded at selvages of these bodies, while hemicrystalline one, in their central parts. These rocks are characterized commonly by a lumpy structure; in addition, they are often porous or amygdaloidal. The amygdules are filled with zeolites.

Hyalopilitic tagamite comprises elongated (0.6–1.2 mm) plagioclase laths, which form stellate accumulations in places (Fig. 38). Hemicrystalline tagamite is characterized by shorter (0.05–0.1 mm) plagioclase laths and isometric pyroxene grains. The latter also forms reaction rims around inclusions. Locally, holocrystalline texture is recorded in larger tagamite bodies.

Holocrystalline tagamites are composed of plagioclase laths up to 0.02 × 0.05 mm in size (25–35 vol.%), isometric and prismatic pigeonite grains up to 0.1 × 0.6 mm (15–25%) and xenomorphic grains up to 0.2 mm across of quartz and potash feldspar, which fill interstices (Fig. 39). Dust-like and well-faced magnetite crystals up to 0.03 mm across contribute up to several vol.%. Chemical compositions of pigeonite, plagioclase (andesine-labradorite), and alkali feldspar are given in Tables 7 and 8. Interstitial brownish glass is commonly replaced by green saponite, chlorite or hydromicas. The clastic inclusions are rare (<3 vol.%); these are represented by recrystallized diaplectic quartz and plagioclase up to 2 mm across

In all types of tagamites, dust-like ore minerals are irregularly scattered throughout the groundmass; pyrite is also rather common. The inclusions (from 1 mm to 1 cm across) are represented by fragments of rocks and minerals. They are frequently surrounded by rims of opaque glass with prismatic pyroxene grains. The clasts of diaplectic orthoclase and diaplectic feldspar glass are recrystallized to radial-fibrous

Fig. 39 Tagamite with holocrystalline ophitic texture. Groundmass is composed of pigeonite poikilocrysts, plagioclase laths, and crystals of magnetite. Clastic inclusions are rare. VDW, depth 1144 m. Photomicrograph, plane-polarized light

aggregates of potash feldspar. Granulated quartz derived from recrystallization of dialpectic quartz glass is common.

Within polymict allogenic breccia, tagamites occur locally at the southwestern slope of the central uplift. The vertical extension of the tagamite bodies there ranges from 11 m (bh. 44) to 30 m (bh. 45). Marginal parts of these bodies are porous and slag-like and often contain a large amount (up to 40 vol.%) of inclusions of fusion gneisses, argillites and sandstones. The inclusions range in size from a millimeter to a few of centimeters, occasionally up to 15 cm. In central parts of the bodies the inclusions amount to 15% at the maximum.

Tagamites from central parts of bodies in allogenic breccia are aphanitic; they show massive structure and hemicrystalline or microdolerite texture (Fig. 40), which combines with a poikylophytic texture in places. The groundmass is composed mainly of plagioclase (andesine to labradorite) laths (up to 0.15 mm across) and hypersthene prisms (from 0.05 to 0.2 mm) (Table 7). Orthoclase, albite and quartz occur in interstices (Table 8).

The differences in composition of pyroxenes from tagamite bodies differing in either size or position (within either authigenic or allogenic breccias) most likely result from different cooling and hence crystallization rates. Besides, pyroxenes from tagamites differ significantly from both pyroxenes of basement crystalline rocks and thermally-metamorphosed shocked rocks (coptoblastoliths, see Sect. 4.2) and from pyroxenes arisen due to hydrothermal alteration (see Fig. 18 in Chap. 4).

Inclusions in hemicrystalline tagamites are represented by mosaic quartz, partly fused plagioclase and augite. Occasionally, pyroxene reaction rims are observed around quartz inclusions. Alteration saponite, less often chlorite and hydromicas, plus pyrite replace the interstitial glass. Rounded amygdales are usually filled with

Table 7 Selected electron microprobe analyses of plagioclase and pyroxene microlites from tagamites

Components	1	2	3	4	5	6	7	8	9	10	11	12
SiO_2	54.68	56.37	56.91	54.29	54.60	55.70	56.78	58.41	57.18	51.39	52.85	53.92
TiO_2	–	–	–	–	–	–	–	–	0.37	0.37	0.61	0.47
Al_2O_3	28.14	26.60	26.10	27.61	26.58	26.74	25.65	25.02	1.41	2.22	2.25	0.62
Fe_2O_3	–	–	–	0.67	0.88	1.95	0.40	1.02	–	–	–	–
FeO	–	–	–	–	–	–	–	–	20.23	24.36	18.59	15.82
MnO	–	–	–	–	–	–	–	–	0.52	0.36	0.30	0.51
MgO	–	–	–	–	–	–	–	–	22.79	18.12	23.33	24.5
CaO	10.82	9.49	10.51	12.04	11.12	10.96	9.31	9.29	1.21	1.81	1.68	3.57
Na_2O	4.92	6.11	5.20	4.41	4.70	5.24	6.01	6.19	–	–	–	–
K_2O	0.61	0.58	0.50	0.24	0.35	0.40	0.56	0.42	–	–	–	–
Total	99.17	99.15	99.22	99.26	98.33	100.99	98.71	100.35	99.71	98.83	99.61	99.41
n	2	1	2	3	3	3	3	1	5	2	2	4
Fs									33	42	30	25
En									65	54	66	68
An, Wo	55	46	53	60	57	54	46	45	2	4	3	7

Electron microprobe analyses, VSEGEI Lab, analyst L. S. Bogomolnaya. *n* is number of repeated analyses
Samples 1 to 8—plagioclase, 9 to 11—hypersthene, 12—pigeonite
Samples 1–5, 8–11 are from borehole 44, depth 267–278 m. Samples 6, 7, 12 are from VDW, depth 1144 m

Table 8 Selected electron microprobe analyses of interstitial glass and microlites of alkali feldspars and albite from tagamites

Mineral	Interstitial glass		K-Na feldspar				Albite		
Borehole	44	44	VDW	VDW	VDW	VDW	VDW	VDW	VDW
Depth	273	273	1144	2410	2410	2568	1144	2410	2472
SiO_2	72.44	72.66	67.82	66.30	66.34	64.10	66.65	70.37	64.53
TiO_2	1.72	0.56	–	0.35	–	0.43	–	–	–
Al_2O_3	11.85	12.50	17.63	18.44	19.27	18.64	21.21	20.68	20.98
FeO	2.17	1.10	–	0.44	–	–	0.56	–	–
MnO	0.32	–	–	–	–	–	–	–	–
MgO	0.16	0.12	0.11	0.26	–	–	–	–	–
CaO	0.97	0.39	0.42	0.45	0.57	0.38	3.20	0.46	2.35
Na_2O	3.49	1.93	3.40	3.08	8.05	–	9.10	9.20	10.91
K_2O	4.45	8.24	11.21	11.39	5.03	16.16	–	0.26	0.36
Total	97.57	97.50	100.59	100.71	99.26	99.71	100.62	100.97	99.13

Electron microprobe analyses, VSEGEI Lab, analyst L. S. Bogomolnaya. *n* is number of repeated analyses

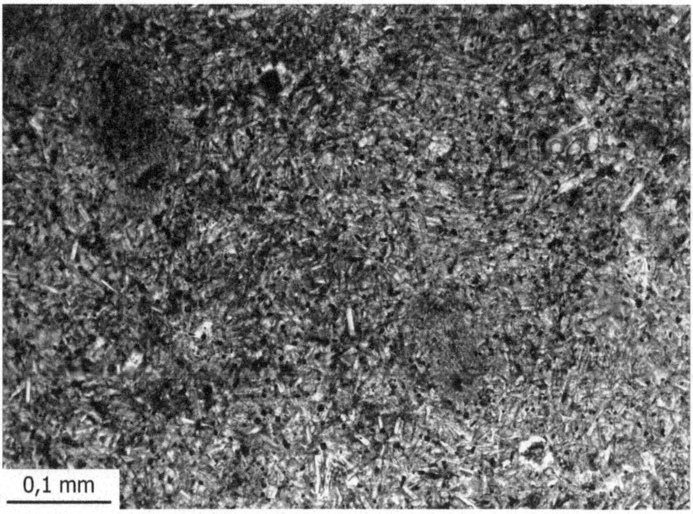

Fig. 40 Tagamite with microdolerite texture of the groundmass consisting of fine (<0.02 mm) plagioclase laths and hypersthene prismatic crystals. Tiny magnetite grains are dispersed. Borehole 44, depth 276 m. Photomicrograph, plane-polarized light

smectites and, in rare places, zeolites. When compared with the bulk composition of tagamites, fresh interstitial glass is enriched in silica and alkalis, especially potassium, and depleted in alumina, magnesium, calcium and iron (Table 8). These differences correspond to general compositional trend of crystallization fractionation of impact melt (Selivanovskaya 1987).

The study of residues of the thermochemical decomposition of tagamites and suevites showed the presence of shock-metamorphosed graphite inclusions as well as of small grains of impact diamonds. The latter were firstly discovered under supervision of O. S. Bogatyrev (Central Geological and Geophysical Expedition of the Industrial Geological Enterprise "Centergeologiya"). Some samples passed by O. S. Bogatyrev to VSEGEI for examination were treated by I. G. Fedorova and briefly described by A. I. Raikhlin and G. I. Shafranovsky (Masaitis and Pevzner 1999). The morphological and spectroscopic characteristics of the diamonds show that they originate from graphite crystals subjected to a high pressure transformation. In general, they are similar to diamonds from the Popigai impact crater (Masaitis et al. 1998; Masaitis 2013; Shafranovsky 2019). The diamonds occur as thin flakes or fragments from about 0.1–0.15 mm across. Elongated or flattened fragments are predominant. Some diamonds are irregularly shaped, some others are paracrystals, which inherite the graphite facets.

Two groups of diamond grains are distinguished. The first group includes colourless, yellow, white, and light brown diamonds, which contribute about 80% of all studied grains. Diamonds of the second group are black, grey, or dark brown. The dark tints are caused by variable amounts of graphite inclusions. The impact diamonds from tagamites and suevites have a corroded surface caused by oxidation by means of the host silicate melt. There are also transparent diamond grains with smooth faces, which apparently did not undergo oxidation; they are characteristic of protoimpactites. The specific weight of the impact diamonds ranges from 3.3598 up to 3.5571 g/cm^3. The X-ray analysis of the diamonds showed that these are polycrystalline aggregates consisting of two phases—the cubic diamond (80%) and the hexagonal lonsdaleite (20%). The ratio of these phases is variable in grains of different colours; the maximum lonsdaleite content has been recorded in colourless grains.

The average content of the impact diamonds in impactites is 0.3 carat per ton. Taken into account their distribution, preliminary estimation of total reserves of the impact diamonds in the area of the Puchezh-Katunki structure may be treated as a mineral occurrence or a small deposit.

The average whole-rock compositions of tagamites from different locations are given in Table 9. They are similar to suevite composition (Table 6) in general. Some differences are caused by the presence of fragments of terrigenous and carbonate rocks in suevites. Chemical variations within either tagamite or suevite populations that depend on sample location, are much more significant than variations between these two groups of impactites. Tagamites from authigenic breccia and from allogenic breccia differ insignificantly in main elements contents; higher volatile content and Fe_2O_3/FeO ratio in the latter are mainly related to more intense post-impact hydrothermal alteration. At the same time, higher values of Zn, Ag, and Cu in the latter tagamites, and of Ba and Rb, in the former one (Table 10), are evident indicating possibly to some compositional variations of melted protoliths and thereby, to incomplete homogenization in the melt zone.

When compositions of tagamites and different crystalline basement lithologies are plotted to the AFM ternary diagram, tagamite points form a relatively compact cluster within the field of basement rocks (Fig. 41). At the same time, some deficit in silica and sodium and excess in calcium, magnesium, and iron in tagamites with respect to the average crystalline target rock composition, indicate unambiguously to any participation of sedimentary cover lithologies in the melt zone; however, these lithologies contribute no more than 5–7%. If this participation having been ignored, products of the solidification of impact melt (tagamites and glass fragments in suevites and coptomict gritstones) generally correspond in their chemical composition to a mixture of various gneisses (leucocratic amphibole-biotite, mesocratic pyroxene-biotite, amphibole, and high-alumina varieties) and amphibolites in the ratio about 3:1. The same is resulted from data on trace elements contents in impact melt and target rocks (Table 10). This is consistent with the estimate of the relative distribution of different crystalline rocks in the upper part of the VDW section. Tagamites, which correspond in composition to the most acidic (granites) or the most basic (ultramafics) lithologies, have not been established. Thus, the melted matter was homogenized to a certain extent, although the mixing remained incomplete.

The compositional compliance of tagamites with crystalline target rocks is confirmed by data on strontium and helium isotope studies (Yushko et al. 1998). The $^{87}Sr/^{86}Sr$ ratio in vein tagamites from authigenic breccia ranges from 0.705 to 0.709 (and it is the same in crystalline target rocks), while for tagamites from allogenic

Table 9 Average composition (wt%) of tagamites from different locations in the Puchezh-Katunki impact structure

Components	1		2		3		4		5	
	X	S	X	S	X	S	X	S	X	S
SiO_2	57.58	1.83	56.61	3.72	55.93	2.93	54.77	3.18	56.49	2.68
TiO_2	0.79	0.08	0.97	0.36	0.98	0.12	0.90	0.21	0.86	0.18
Al_2O_3	15.24	1.07	15.43	0.72	15.86	1.76	15.04	1.64	15.24	1.20
Fe_2O_3	3.47	0.70	3.31	0.98	5.08	2.11	5.81	1.57	4.12	1.45
FeO	3.62	0.68	3.79	1.00	3.82	2.08	3.05	1.80	3.60	1.28
MnO	0.12	0.02	0.09	0.03	0.12	0.07	0.20	0.10	0.14	0.07
MgO	4.56	0.65	3.89	0.89	3.69	0.40	4.69	0.86	4.44	0.77
CaO	5.73	0.94	4.52	2.53	6.09	0.90	4.97	1.00	5.46	1.29
Na_2O	4.45	0.50	2.21	1.20	3.53	0.67	2.89	0.44	3.70	1.04
K_2O	1.76	0.23	4.68	2.54	1.59	0.54	0.74	0.70	1.85	1.45
P_2O_5	0.18	0.05	0.14	0.06	0.26	0.14	0.08	0.09	0.16	0.10
Cr_2O_3	0.03	0.02	0.02	0.03	–	–	0.01	0.01	0.02	0.02
L. O. I.	2.57	1.25	4.13	1.29	3.27	1.13	6.33	3.78	3.83	2.70
Total	100.10	–	99.79	–	100.22	–	99.48	–	99.91	–
n	17		4		5		9		35	

X Mean value; *S* Standard deviation

XRF data, analyzed by B. A. T simoshenko, VSEGEI Labs, St. Petersburg, with a SRM-2 instrument. *n* is number of samples analyzed

1 to 3—Tagamites from veins cut the authigenic breccia: 1, 2—VDW core at depths above 3 km (1) and below 3 km (2); 3—from boreholes drilled in the crest and slopes of the central uplift; 4—Tagamites from irregularly-shaped bodies in the allogenic breccia on the southwest slope of the central uplift; 5—averaged of 1–4

Table 10 Average contents of selected trace elements (ppm) in different lithologies from the Puchezh-Katunki impact structure

Lithology	Coptomict gritstone	Suevite	Tagamite from authigenic breccia	Tagamite from allogenic breccia	Various gneisses	Amphibolite and schist	Granite gneiss and granite
Ag	<0.1	0.16	0.22	0.36	0.13	0.14	0.14
Co	24	19	25	23	17	47	5.9
Cu	38	32	35	54	22	58	42
Ni	44	41	50	38	27	76	9.2
Zn	66	79	94	140	77	131	85
Cr	102	88	118	91	57	134	20
V	107	87	116	120	74	225	21
Sr	1768	1179	374	429	375	208	384
Y	12	12	17	21	8.8	18	4
Zr	89	107	157	172	97	69	117
Nb	9.4	9.2	14	15	8.5	9.7	7.5
La	28	30	42	36	24	16	26
Ce	55	59	78	68	47	35	48
Be	n.d.	n.d.	1.1	1.4	0.89	0.90	1.1
Ba	n.d.	n.d.	749	525	567	333	1577
Pb	n.d.	n.d.	9.5	11	7.0	3.3	16
Rb	n.d.	n.d.	45	8.6	34	14	78
n	3	3	7	5	36	17	11

All measurements are performed in VSEGEI Labs.. Ni, Co, Cu, Ag, Cr, V, and Zn are determined by ICP AES (OPTIMA-4300 instrument, analyst E.G. T chervyakova); Ba, Be, ad REE, by ICP-MS (ELAN–DRC-6100, analyst V. A. Shishlov); Rb, Sr. Pb, Y, Zr, and Nb, by XRF (ARF-6, analyst L. A. Matveeva)
n number of samples analyzed

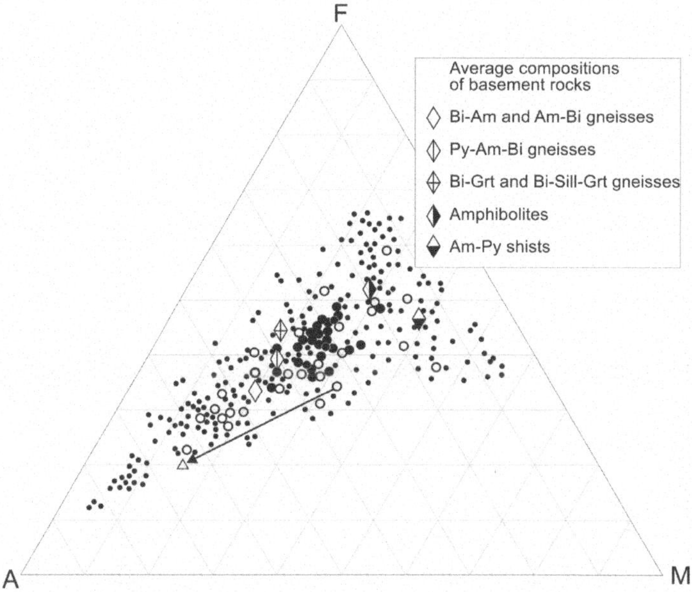

Fig. 41 AFM ternary compositional diagram for tagamites in comparison with rocks composing the upper part of the central uplift (down to 3080 m deep). Individual analyses of basement metamorphic rocks are plotted as small black circles, tagamites, as large black circles, coptoblastoliths, as open circles. A granophyre from a depth of 1380 m (see Sect. 4.2) in the VDW is shown by an open triangle, which is connected by an arrow with host coptoblastolith

breccia, from 0.711 to 0.713. This difference may be explained by some contribution of Neoproterozoic argillites, in which the $^{87}Sr/^{86}Sr$ ratio varies from 0.720 to 0.738, to the melted substrate for the latter. Helium isotope ratio ($^{3}He/^{4}He$) in tagamites from both positions ($2.1–4.7 \times 10^{-8}$ from authigenic breccia and $1.9–11.7 \times 10^{-8}$ from allogenic breccia) is the same that in the target rocks ($2.5–4.8 \times 10^{-8}$) indicating the absence of any abyssal matter input (Yushko et al. 1998). Both Sr and He isotope compositions are typical for the upper continental crust.

Rare earth concentrations in all tagamites vary insignificantly (Table 11). At the same time, chondrite-normalized REE abundances of tagamites show that impact melt bodies within allogenic breccia are more completely homogenized than those making up veins in authigenic breccia (Fig. 42). In suevites and coptomict deposits REE concentrations are essentially (for 10–30%) lower due to, more probably, an addition of a sedimentary component. However, REE abundances of suevites show unexpectedly a higher uniformity compared with tagamites; this confirms an inhomogeneity of melted substrate for different tagamite bodies.

A comparison of REE contents in crystalline target rocks and impactites shows that tagamites and gneisses are similar by the REE distribution patterns (Fig. 42); all these manifest a weak negative Eu anomaly characteristic for formations composing the upper continental crust (Taylor and McLennan 1985). However, a clear enrichment of

Table 11 Average REE content (ppm) in different lithologies from the Puchezh-Katunki impact structure

Lithology	Coptomict gravelite	Suevite	Tagamite from authigenic breccia	Tagamite from allogenic breccia	Various gneisses	Amphibolite and schist	Granite and granitic gneiss
La	27.77	30.13	43.70	35.90	36.97	22.16	35.76
Ce	55.13	60.47	81.16	68.64	70.80	49.05	67.07
Pr	6.12	6.82	8.95	8.00	7.70	6.11	7.00
Nd	22.10	24.27	32.33	28.40	27.18	24.53	23.54
Sm	4.00	4.21	5.43	5.46	4.56	5.23	3.62
Eu	1.04	1.07	1.36	1.36	1.20	1.55	1.14
Gd	3.72	3.96	4.52	5.00	3.60	4.68	2.64
Tb	0.55	0.57	0.62	0.70	0.49	0.78	0.35
Dy	3.05	3.02	3.38	4.26	2.57	4.48	1.86
Ho	0.57	0.55	0.70	0.82	0.47	0.86	0.34
Er	1.60	1.50	2.11	2.35	1.29	2.38	0.99
Tm	0.25	0.23	0.33	0.37	0.18	0.33	0.15
Yb	1.65	1.55	2.19	2.30	1.23	2.18	0.97
Lu	0.25	0.24	0.33	0.37	0.19	0.33	0.17
Eu*	0.83	0.81	0.85	0.79	0.98	0.94	1.63
ΣREE	127.79	138.59	187.11	163.92	158.42	125.88	145.60
LREE/HREE	8.90	10.43	10.77	8.09	20.77	7.65	26.34
n	3	3	7	5	34	17	14

REE are determined by ICP-MS, with ELAN–DRC-6100 instrument, VSEGEI Lab., V. A. Shishlov, analyst. *n* is number of samples measured

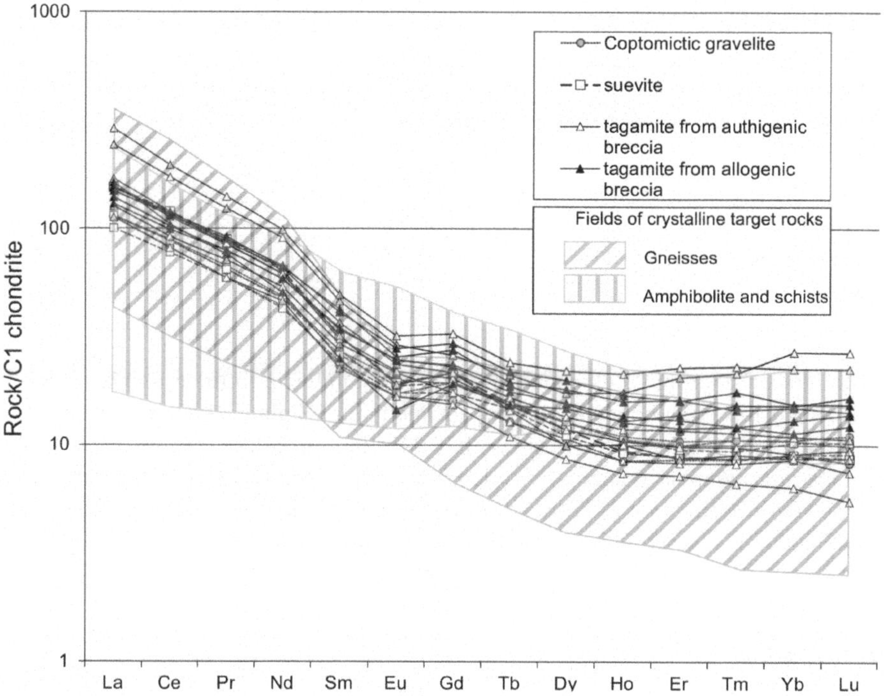

Fig. 42 Chondrite-normalized (Taylor and Maclennan 1985) REE patterns of impactites and coptomict deposits from the Puchezh-Katunki impact structure as compared to main types of crystalline target rocks

impact melt in REE is revealed (Table 11); additional input of REE into impact melt is estimated to range from 10 to 30%; it is the most significant for HREE (Fig. 42). Concentrations of other incompatible elements in impact melt rocks are also higher compared to the target rocks (Table 10). This may indicate a selective melting of plagioclase-enriched fraction of substrate.

In order to detect a probable projectile matter in impactites, thirty instrumental neutron-activation analyses (INAA) of cores from VDW and other boreholes were performed. Some siderophile and platinum group elements concentrations, measured at the Vernadsky Institute (Russian Academy of Sciences) by K. A. Lorentz under the supervision of M. A. Nazarov, are given in Table 12. Among the analyzed samples are gneisses, schists, amphibolites, mafic and ultramafic rocks of the crystalline basement (9 samples), various allogenic breccias, suevites and tagamites of the crater fill (15 samples), coptomict gritstones and sandstones of the crater lake deposits (6 samples). The results showed that impactites, particularly suevites, are enriched by factor 3.9–1.2 by some elements typical for cosmic matter, especially Os and Ir, with respect to the 3:1 gneiss/amphibolite mixture. Taking into account that the REE input into impact melt from melanocratic mafics and ultramafics is most likely negligible due to insignificant contribution of these lithologies to the crystalline basement,

Table 12 Contents of some siderophile elements and PGE in crystalline target rocks and impactites (INAA results)

Lithology	Cr (ppm)	Co (ppm)	Os (ppb)	Ir (ppb)
Mixture of gneisses and amphibolites (with 3:1 ratio)	37	18.5	0.15	0.12
Tagamites	129	20	0.18	0.13
Suevites	93	22	0.31	0.28
Tagamites (increase factor)	3.5	1.08	1.20	1.23
Suevites (increase factor)	2.5	1.19	2.07	2.43

the excess of siderophile elements and PGE may indicate to the possible presence of a meteoritic component in the impact melt (e.g., Palme 1982; Feldman et al. 1984; Schmidt et al. 2018). However, ICP-MS data for a larger sample (Table 10) demonstrate identical contents of siderophiles (Co, Ni, and Cr) in tagamites and averaged crystalline target. Thus, a limited number of analyses and discrepancies in the obtained results require to continue these studies.

References

Bates RL, Jackson JA (1987) Glossary of geology, 3rd edn. American Geosciences Institute, Virginia, p 788

Blom GI (1960) On age of the conglomerate-clay sequence of Kerzhenets, Linda, Vedomost', and Serga rivers basins (in Russian). Sov Geol 4:118–120

Bogorodskaya OA, Tumanov RR (1980) Geological map of the USSR on 1:200,000 scale. Sheet O-38-XXVI: explanatory notes. Moscow (in Russian)

Chao ECT, Xie XD (1990) Mineralogical approaches to geological investigations. Science Press, Beijing, p 388

Feldman VI, Sazonova LV, Nosova AA (1984) Geological structure and petrography of impactites of the Puchezh-Katunki astrobleme (Volga Region). Bull Mosc Soc Nat Res Sect Geol 6:53–63 (in Russian)

Masaitis VL (2005) Morphological, structural and lithological records of terrestrial impacts: an overview. Aust J Earth Sci 52:509–528

Masaitis VL (2013) Impact diamonds of the Popigai astrobleme: main properties and practical use. Geol Ore Depos 55(8):607–612

Masaitis VL (ed) (2019) Popigai impact structure and its diamond-bearing rocks. In: Impact studies. Springer, 205pp, https://doi.org/10.1007/978-3-319-77988-1

Masaitis VL, Naumov MV (1990) Structural observations in the west slope of the Puchezh-Katunki astrobleme (abs.). In: Abstracts for 21st all-union meteoritic conference. Moscow, pp 135–136 (in Russian)

Masaitis VL, Pevzner LA (eds) (1999) Deep drilling in the Puchezh-Katunki impact structure. VSEGEI Press, St. Petersburg, p 392 (in Russian)

Masaitis VL, Danilin AN, Mashchak MS, Raikhlin AI, Selivanovskaya TV, Shadenkov EM (1980) Geology of astroblemes. Nedra Press, Leningrad, p 231 (in Russian)

Masaitis VL, Mashchak MS, Naumov MV Selivanovskaya TV, Orlova JV (1995) The Puchezh-Katunki impact crater: main features of geological structure (in Russian) Trans Dokl Russ Acad Sci 342:358–360

Masaitis VL, Mashchak MS, Raikhlin AI, Selivanovskaya TV, Shafranovsky GI (1998) Diamond-bearing impactites of the Popigai astrobleme. VSEGEI Press, St. Petersburg, p 178 (in Russian)
Masaitis VL, Petrov OV, Naumov MV (2018) Impact lithologies—a key for reconstruction of rock-forming processes and a geological model of the Popigai crater, northern Siberia. Aust J Earth Sci. https://doi.org/10.1080/08120099.2018.1509372
Murawsky H (1977) Geologisches Wörterbuch. F. Enke Pub., Stuttgart, p 280
Naumov MV (2002) Impact-generated hydrothermal systems: data from Popigai, Kara, and Puchezh-Katunki impact structures. In: Plado J, Pesonen LJ (eds) Meteorite impact structures in precambrian shields. Impact studies, vol 2, Springer, Berlin, pp 117–171
Palme H (1982) Identification of projectiles of large terrestrial impact craters and some implications for the interpretation of Ir-rich cretaceous/tertiary boundary layers. In: Silver LT, Schultz PH (eds) Geological implications of impacts of large asteroids and comets on the earth. Geological Society of America. Special Paper 190, pp 223–233
Schmidt G, El Goresy A, Palme H (2018). Extraterrestrial platinum group elements in impactites and misleading Cr/Ni, Ni/Co and Cr/Co element ratios for projectile identification (abs.). In: 81st annual meeting of the meteoritical society. LPI Contrib. No. 2067, #6073, Moscow
Selivanovskaya TV (1987) Crystallization fractionation of impact melts. Meteoritika 46:128–135 (in Russian)
Selivanovskaya TV (1992) Impactites of Puchezh-Katunki astrobleme (abs.). In: Abstracts for 16th Vernadsky-Brown Microsymposium. Vernadsky Institute, Moscow, pp 75–76
Selivanovskaya TV, Mashchak MS, Masaitis VL (1990) Impact breccias and impactites of Kara and Ust-Kara astroblemes. In; Masaitis VL (ed) Impact craters on Mesozoic-Cenozoic boundary. Nauka Press, Leningrad, pp 55–96
Shafranovsky GI (2019) Impact diamonds from shocked crystalline rocks and impactites. In: Masaitis VL (ed) Popigai impact structure and its diamond-bearing rocks. Impact studies. Springer, pp 137–156
Taylor SR, McLennan SM (1985) The continental crust, its composition and evolution. Blackwell Scientific, Oxford, p 312
Tikhomirova GA (1967) Map of anomalous magnetic field of the USSR of 1:200,000 scale. Sheets O-38-XXVI, XXVII, XXXII, XXXIII. VSEGEI, Leningrad (in Russian)
Tumanov RR, Bogorodskaya OA (1975) Geological map of the USSR on 1:200,000 scale. Sheet O-38-XXVI. VSEGEI, Leningrad (in Russian)
Yushko NA, Kremenetsky AA, Vorontsov AK (1998) Geochemical features of shock metamorphic process: evidences from impact melt rocks (based on results of deep drilling in the Puchezh-Katunki astrobleme). Geochlimija 7:656–667 (in Russian)
Zander VN, Tomashunas YuI, Berkovsky AN, Suvorova LV, Dedeev VA, Kratz KO (1967) Geological structure of the basement of Russian platform. Nedra Press, Moscow, p 124 (in Russian)

Transformations of Crystalline Rocks of the Central Uplift and Its Deep-Seated Zones

Victor L. Masaitis, Jeanne V. Orlova, Mikhail V. Naumov and Mikhail S. Mashchak

1 Shock Metamorphism

Various mineralogical and petrographic features of the central uplift rock alteration have been studied in the cores of VDW and in other wells, which allowed to estimate main parameters of various processes of transformation. A few general comments need to be made in advance as to the use of various methods of such assessment, especially the parameters of progressive impact metamorphism. To determine the amplitude of shock compression of rocks and minerals, conventional geobarometers were used (e.g. Stöffler 1971; Harrison and Hörz 1981; Basilevsky et al. 1983; Badyukov 1986; Langenhorst and Deutsch 1998; French 1998; French and Koeberl 2010). Various observed deformational features in crystals (dislocations, planar fractures, planar deformational features, mechanical twins, kink bands, mosaicism), as well as different kinds of phase transformations of minerals (formation of high-pressure polymorphs, diaplectic glass, decomposition and melting and evaporation) may be applied in this purpose. The most reliable estimation of the impact compression amplitude may be considered based on of the development of planar deformation features (PDF) in quartz and its phase transitions in non-porous rocks (Langenhorst and Deutsch 1998; Stöffler and Langenhorst 1994; etc.). In addition to the amplitude of the shock wave, it is necessary to bear in mind the influence on these transformations of some other factors, including the initial rock temperature, duration of the compression pulse, reverberation of compression due to reflections, orientation of anisotropic crystals in relation to the shock front and their size, acoustic impedance of the surrounding matrix, the presence of water vapors, etc.

Estimation of the shock compression from the PDF statistics in quartz is reasonable within the limits of quite powerful (many hundreds of meters) concentric zones, on the basis of averaging data on numerous rock samples selected in the massif along

V. L. Masaitis (✉) · J. V. Orlova · M. V. Naumov · M. S. Mashchak
A.P. Karpinsky Russian Geological Research Institute,
199106 Sredniy prospekt 74, St. Petersburg, Russia
e-mail: vcmsts@mail.ru; Victor_Masaitis@vsegei.ru

V. L Masaitis and M. V Naumov (eds.), *The Puchezh-Katunki Impact Crater*,
Impact Studies, https://doi.org/10.1007/978-3-030-32043-0_4

the direction of shock wave propagation, as well as taking into account changes in a number of rock-forming minerals. It is necessary to observe the requirements of comparability of local estimates of impact compression amplitude by diaplectic changes with its real gradient on a kilometer scale, neglecting insignificant fluctuations caused by the above-mentioned factors.

To estimate the variability of shock compression by depth and laterally along the radii from the geometric center of the Puchezh-Katunki structure, the maxima of the obtained values are used, based on the critical changes of minerals or newly formed phases, established by petrographic methods, which is a criterion for drawing conditional boundaries between such zones differing in the level of shock-metamorphism. The following shock effects are used as critical: first appearance of PDFs with the orientation $\{10\bar{1}3\}$ (ω-system) corresponds to a minimum pressure of 10 GPa, and the orientation $\{10\bar{1}2\}$ (π-system) to a pressure in excess to 20 GPa. The occurrence of stishovite indicates a minimum shock pressure of 12–15 GPa, of coesite, of more than 30 GPa, of diaplectic glass, of more than 35 GPa, while shock melting of quartz to yield lechatelierite needs post-shock temperatures that correspond to a shock pressure of more than 50 GPa (e.g., Stöffler and Langenhorst 1994, and references therein).

Other phase transformations, which reliably indicate the level of shock metamorphism, are appearance of maskelynite, i.e. diaplectic plagioclase An_{15-35} glass, that indicates shock pressure more than 30 GPa, and formation of plagioclase fusion glass (more than 45 GPa). Conversion of graphite into diamond-lonsdaleite polycrystalline aggregate occurs at shock pressures of 35–45 GPa. Full melting of crystalline rocks takes place at shock loading of about 60 GPa.

An important petrographic indicator correlating with the shock pressure is reduced birefringence of tectosilicates, which is distinguishable at initial stages of shock transformations. It is amplified up to the complete isotropization of minerals with increasing pressure (formation of diaplectic glasses).

A general scheme of radial and concentric zoning of shock transformations based on the mentioned criteria is presented in Table 1 and in Figs. 1 and 2.

Due to superposed post-shock thermal and hydrothermal alterations, shock effects can be significantly obliterated or modified. The barometric estimates can be made on the reconstructed primary traces of compression based on ontogenetic interrelations between new-formed mineral phases. This reconstruction is facilitated by residual post-shock heating and non-monotonous character of superimposed thermal metamorphism. Among its typical features are abrupt temperature fluctuations and caused by them transformations in narrow spatial limits, and besides on the background of rocks of similar composition that have experienced the same shock load.

Shock-metamorphosed changes are established in rocks and minerals for a distance of about 5 km from the Puchezh-Katunki crater center and are traced along the VDW core down to its bottom, i.e. for more than 4.2 km from the top of the central uplift (Fig. 3).

In the VDW, which locates approximately 1500 m northeast of the geometric center of the crater, fluidal quartz fusion glass (lechatelierite) is found in shocked biotite-amphibole gneisses just beneath of the base of allogenic breccia. Plagioclase is

Table 1 Principal mineralogical criteria and zoning of shock metamorphism in the Puchezh-Katunki impact structure

Depths in the VDW (H, in m)	Radial distance from the crater center (R, in m)	Critical impact transformations of tectosilicates	Shock pressure estimate, GPa
550–600	550–1300	Plagioclase → fusion glass and diaplectic glass (maskelynite) Quartz → diaplectic glass, in places lechatelierite	50–45
600–1800	1300–3400	Plagioclase → diaplectic glass (maskelynite) Quartz → diaplectic glass, in places coesite Graphite → diamond	45–35
1800–3300	3400–5000	Quartz →quartz with PDF (3–5 systems; ω and π s prevail, ξ, r, s, and ρ are common)	35–25
3300–4500	?	Quartz → quartz with PDF (3–5 systems; ω and π prevail, ξ and r are common)	25–20
4500–5374	?	Quartz → quartz with PDF (1–3 systems; c and ω prevail)	20–15

converted into fusion glass, but in some spots maskelynite is preserved. Amphibole is characterized by four systems of planar fractures. One of such systems closely spaced (about 0.01 mm apart) is oriented parallel to the coarse cleavage cracks aligned to (010). Biotite displays several intersecting systems of closely spaced kink bands.

Within a radius of 1300 m from the center of the crater (boreholes 13, 19, 61, 64 etc.), shock fusion plagioclase glasses are observed in gneisses, but the main part of this mineral is transformed into maskelynite. These glasses display fluidal structure or micro-perlitic cracks. In places, the transformation follows only one of the polysynthetic twin systems, which shows the influence of the orientation of shock wave front on transformation of individual crystals. Quartz and microcline are converted into diaplectic glasses; the changes of biotite and amphibole are similar to mentioned above. Samples from boreholes 19, 54, 55 also contain plagioclase fusion glasses and completely amorphized quartz. Due to post-shock annealing, felsic minerals glasses are recrystallized, while mafic minerals are opacitized.

The shock melting of quartz is rarely observed in the uppermost of the authigenic breccia in the VDW section. This makes it possible to assume that in the central zone of the shock compression its amplitude reached 50 GPa. The lower boundary of this

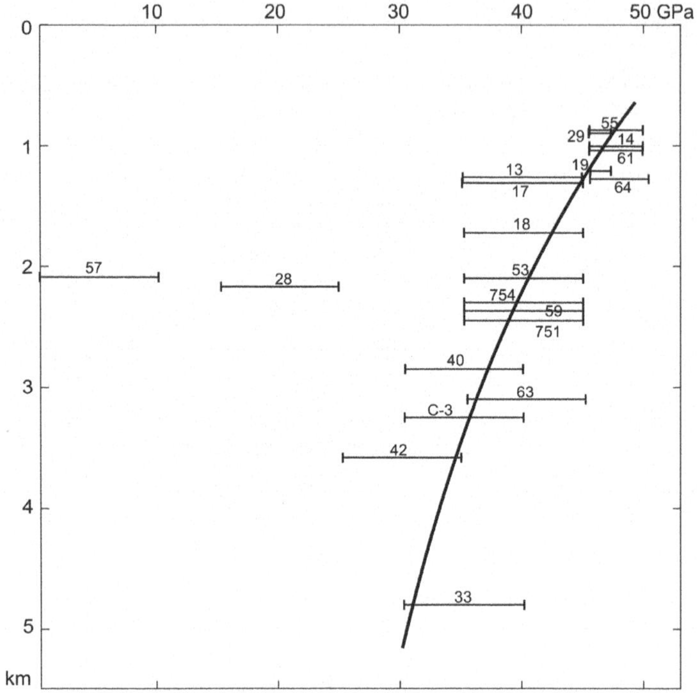

Fig. 1 Estimation of shock pressure in the near-surface zone of the central uplift along radii. Vertical coordinate is distance from the crater center (in km); horizontal coordinate, reconstructed shock compression (in GPa). Horizontal lines show pressure limits estimated from the study of PDF in quartz. Numerals are numbers of boreholes, location of which is given in Fig. 2 in Chap. 3

zone (about 45 GPa) is determined by the coexistence of diaplectic quartz and microcline glasses, as well as plagioclase fusion glasses, which are traced down to a depth of about 600 m. The state of mafic minerals, which underwent decomposition under high post-shock temperatures confirms this assessment. Relatively rapid quenching of shock-heated rocks directly beneath allogenic breccia leads to the preservation of incompletely decomposed biotite and amphibole in some cases.

At the depth from 600 to about 1800 m in the VDW as well as at a distance between 1300 and 3400 m from the crater center in other wells, the shock pressure can be restored only through relics of diaplectic glasses after quartz and feldspars. Quartz diaplectic glasses are usually preserved in the central parts of grains, but at peripheral parts they are recrystallized in quartz with monocrystal structure and palimpsest traces of planar deformation elements. Similar features are observed in boreholes S-3, 18, 40 etc. In places, cryptocrystalline finger-like and kidney-like aggregates of fine-grained quartz are recorded in recrystallized diaplectic glass; these are products of coesite inversion (borehole 13). Both in the depth interval of 1520–1800 m in VDW and in boreholes S-3, 7, 13, 18, and 40, some sections of plagioclase diaplectic glass in feldspar grains with partially recovered by annealing crystalline structure with

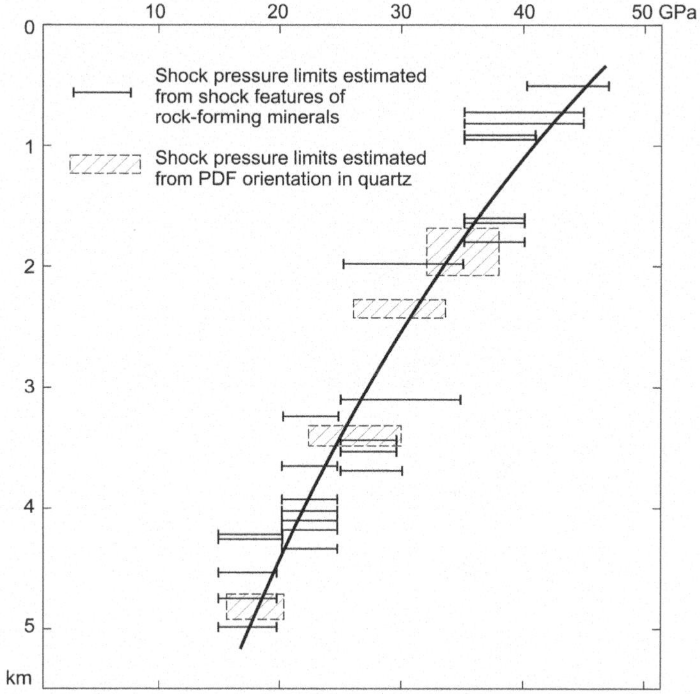

Fig. 2 Estimations of shock pressure values in crystalline target rocks through the VDW section. Vertical coordinate is depth (in km), horizontal coordinate, reconstructed shock compression (in GPa)

cloudy extinction are recorded. In addition, kink bands in biotite (Fig. 4) as well as two or three systems of planar microstructures (including shock twins) and some reduction of birefringence in hornblende occur.

In a sample of biotite-amphibole gneiss from the depth of 607 m in the VDW section, some impact diamonds were found, they were revealed together with numerous graphite scales. This fact indicates a lower limit of shock compression of about 35 GPa. The lack of fusion feldspar glasses constrains the upper limit of shock compression by 45 GPa.

Between 1800 and 3300 m in the VDW section and at a radius of 3400–5000 m laterally (boreholes 40, S-3, 42 etc.), the following shock transformations occur: within annealed quartz with recovered monocrystal texture, sections with lower refractive index and systems of decorated planar elements are preserved (Fig. 5). Both recrystallized and relict domains within quartz grains do not differ in extinction; this indicates the preservation of optical orientation in the recrystallized material. This feature makes it possible to determine the orientation of planar deformation features relative to the optical axis if their traces are present. The average number of relict systems per one grain varies from four (in the depth interval of 1670–2106 m) to three (2257–2420 m). In both cases PDFs orientated along ω- and π-rhombohedrons

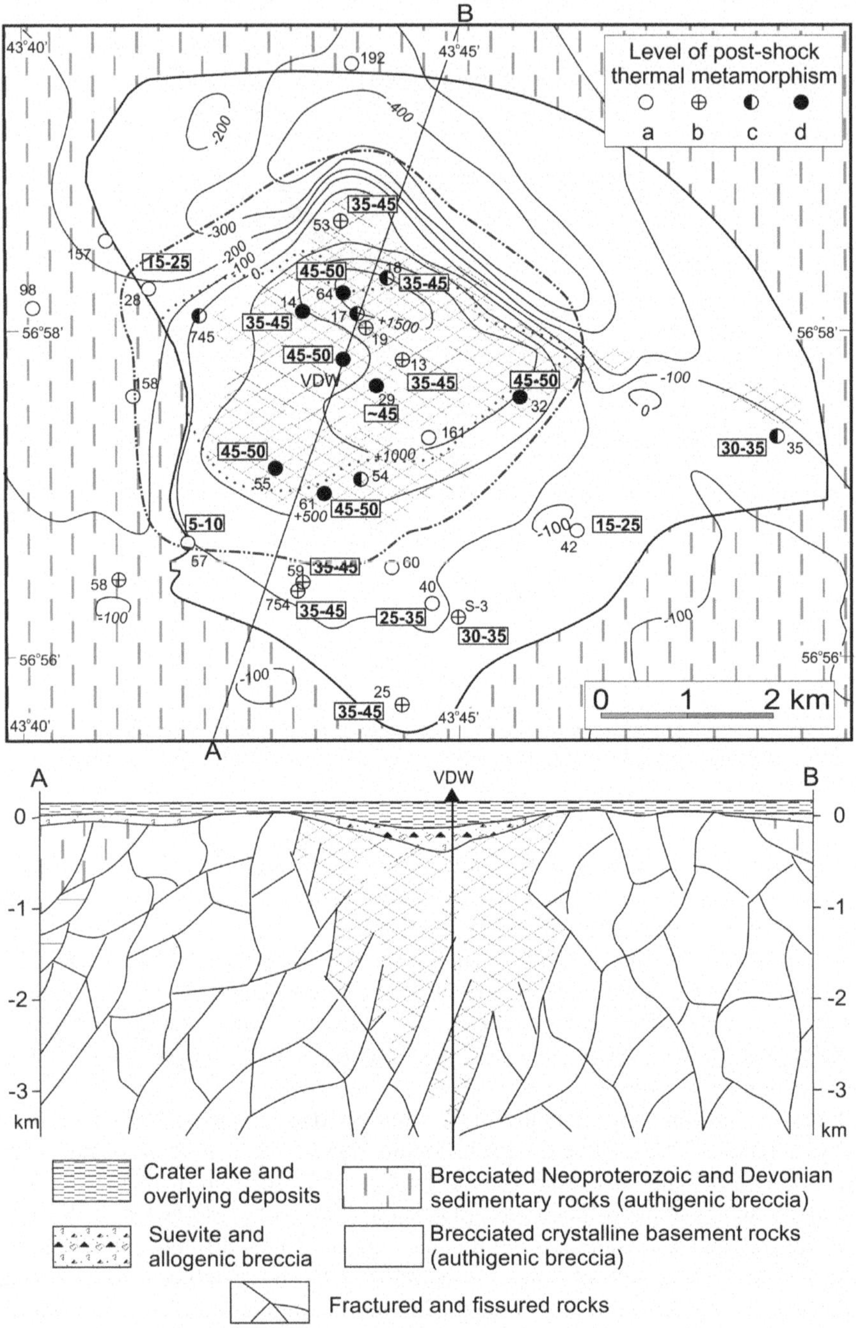
Level of post-shock thermal metamorphism
a b c d
0 1 2 km
A
B
VDW
Crater lake and overlying deposits
Suevite and allogenic breccia
Brecciated Neoproterozoic and Devonian sedimentary rocks (authigenic breccia)
Brecciated crystalline basement rocks (authigenic breccia)
Fractured and fissured rocks

◂**Fig. 3** Distribution of deformed target rocks undergone shock and post-shock thermal metamorphism to a different degree, within the central uplift of the Puchezh-Katunki impact structure. Shock pressure values estimated from mineralogical effects recorded in cores is given for certain numbered boreholes Areas of thermal transformation of crystalline basement rocks are hatched (hatching density reflects the intensity thermal of transformations). In addition, this intensity in the upper part of the authigenic breccia column is shown by differently filled circles for numbered boreholes, in which thermal metamorphic effects were estimated: a—intense, b—moderate, c—weak, d—not appeared. Because of magnetization of rocks is mostly due to post-shock thermal metamorphism, ΔT isogams (in nT) are given. Dot-dashed line shows the crest of the central uplift, dotted line, contour of the central pit. In addition, a section cut across central uplift at A–B line is shown

Fig. 4 A biotite grain with kinkbands (VDW, depth 2001 m) in a shocked biotite gneiss. All quartz grains are characterized by multiple PDFs. Photomicrograph, plane-polarized light

are predominant (Fig. 6). Other PDF planes are also observed, but they are subordinate. It is possible that some of these PDF planes could have been destroyed by the annealing and recrystallization of shocked rocks.

In plagioclase, some of the twin lamellae are converted to maskelynite while others have a reduced birefringence and characteristic intersecting PDE systems that are orientated relative to the twin plane (Fig. 7). In microcline, systems of thin parallel planar fractures oblique to the microcline lattice are recorded. In mafic minerals, deformation feature systems differing by orientation are observed as well. The occurrence in the considered zone of diaplectic quartz with a highly reduced birefringence, maskelynite, and numerous PDE systems in all minerals allows to estimate the shock pressure between 35 and 25 GPa (Fig. 6).

Downward in the VDW section to a depth of about 4500 m, diaplectic transformations in the minerals also occur, but their intensity is decreasing. Relict areas with reduced birefringence occur rarely in quartz grains. The average number of PDFs

Fig. 5 Three systems of planar deformation elements in quartz (VDW, depth 1711 m) in a leucocratic biotite-plagioclase gneiss. Photomicrograph, cross-polarized light

per a quartz grain is three like in the above depth interval, but their total number in the depth range of 3353–3491 m decreases. Predominant PDF orientations are the same—ω- and π-rhombohedrons, but a small peak corresponding to the direction {11$\bar{2}$2} appears on the histogram (Fig. 6). Isotropic feldspars are very rare. Lamellae of multiple twins in a feldspar grain show a various degree of lowering of birefringence and refractive indices. At depths of 3900 m and more, only some twin lamellae in plagioclase (in places, in microcline) have a reduced refractive indice, but systems of planar fractures are preserved (Fig. 7). If to total, the presence of plagioclase with the mentioned optical characteristics, the predominance of ω- and -π orientations of PDFs in quartz, the presence of PDFs in feldspars and biotite (Fig. 8) suggest that the shock pressure at the depth from 3300 down to 4500 m was at least 20 GPa (Table 1).

In crystalline rocks occurring below 4500 m, the total amount of quartz grains with PDF decreases sharply; the average number of PDF systems per grain reduces to be 2. The percentage of PDFs along with ω-rhombohedron increases sharply, while the share of the π-system decreases significantly. Planar fractures, which are parallel to the basal plane {0001} become more abundant. Plagioclase and microcline are characterized by selective reduction of birefringence and, in places, by short fractures oblique to the twin planes. These transformations expressed by a noticeable optical effect are due to the preferred compression of feldspar crystal lattices along the directions [100] and [010] in comparison with [001] (Dworak 1969; Ostertag 1983). For plagioclase, this effect is provided by a shock pressure from 18 to 30 GPa, and for microcline, from 15 to 27 GPa. Since the marked effects in quartz are realized under the shock pressure less than 20 GPa, the shock compression at this depth interval may be estimated at 15–20 GPa.

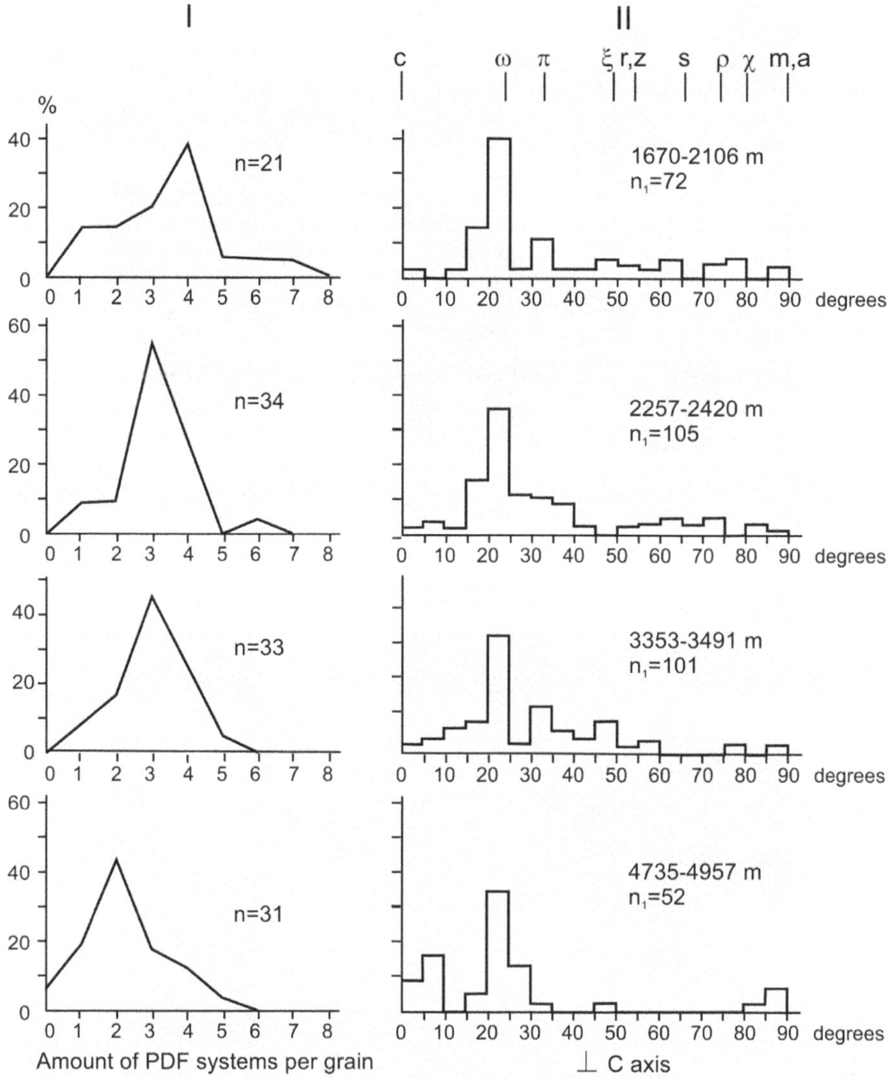

Fig. 6 Number of PDF systems (I) and PDF orientation histograms (II) of PDF in quartz grains for different zones through the VDW section. n and n_1 are number of measured samples. At vertical axis, parts of total number of measured grains in % (I) and measured orientations of PDFs in % (II). On the *x*-axis, the amount of different PDF systems per quartz grain (I), and the angle between the quartz *c*-axis and the pole to the planar feature is plotted. *Y*-axis indicates a percentage of total number of measured grains (I) and frequency for each given angle

Fig. 7 Photomicrograph of the plagioclase, in which one of twins systems transformed into maskelynite with traces of planar deformation (VDW, depth 3781 m). cross-polarized light, ×150

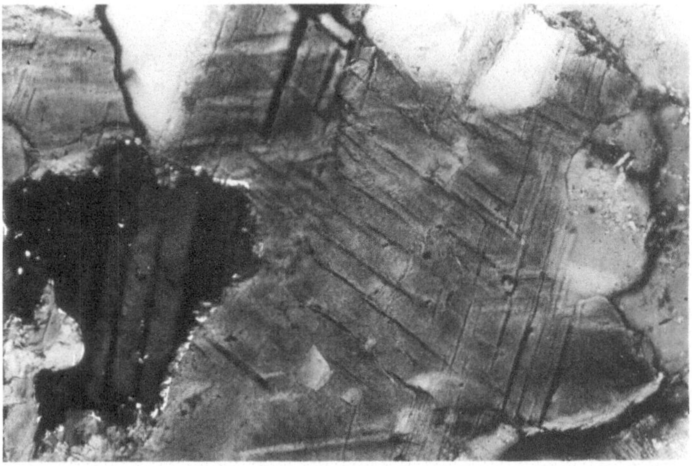

Fig. 8 A biotite scale with two systems of planar deformational bands. VDW, depth 3113 m; the field size is 0.6 mm wide. Photomicrograph, cross-polarized light

Thus, the attenuation of shock deformation of crystalline rocks in the central uplift both downwards and outwards from the center is clearly established. An extrapolation of the shock compression attenuation downward in the section shows that the complete disappearance of shock features is expected at a depth from 8 to 10 km down from the upper boundary of the central uplift. This estimated attenuation of shock effects with depth is well correlated with the decreasing degree of crushing and cataclasis of crystalline rocks, as well as with a gradual increase in their density.

2 Thermal Metamorphism

The thermal metamorphism of crystalline rocks of the central uplift occurred after passing of shock wave. It resulted in recrystallization of minerals, in their decomposition, and in some cases in melting and remobilization. Simultaneously, significant changes in the magnetic properties of the rocks occurred there. As early as during initial core study in the process of deepening of the VDW it was found that thermal transformations of the rocks at many places occurred at temperatures significantly exceeding the post-shock temperatures estimated from the respective levels of shock pressure (Masaitis and Mashchak 1990). These data initiated a detailed study of cores from other boreholes penetrating the central uplift as well as experiments on annealing of shock-metamorphosed rocks from the Puchezh-Katunki impact structure.

The intervals where rocks display intense thermal transformations were found both in the pilot well and in the VDW as well as in many other boreholes penetrating into deformed crystalline rocks of the central uplift (Fig. 3). The areal of intensely thermally-metamorphosed rocks is confined with the area of the central pit, weak and, rarely, moderate thermal transformations are recorded in places in rocks from the crest and outer slope of the central uplift. Inferred from distribution of depth intervals of the most intense thermal metamorphism in VDW and the pilot well, these intervals may be interpreted as intersections of a single zone, which is steeply sinking and splintered with depth. The zone is to be probably several tens of meters thick.

The scheme proposed by Stöffler (1972, 1984) is commonly used for the estimation of post-shock temperatures in non-porous rocks; the temperatures depend mainly on the shock pressure. As it was shown above, the shock pressure in the central uplift of the Puchezh-Katunki exceeded 45 GPa; it is quite probably that it was slightly higher. Hence, the residual temperature there could reach 900 °C at a depth about 550–600 m in the VDW. In the peripheral part of the shock transformation zone where the pressure did not exceed 35 GPa, the post-shock temperature is estimated to about 300 °C (at a depth about 1800 m). Nevertheless, in this zone as well as in the surrounding zone, where the pressure was ~25–35 GPa and the residual temperatures were less than 200–300 °C, various mineralogical features indicate thermal transformations at much higher temperatures than post-shock temperatures given above.

The general scheme of attenuation of post-shock temperature with depth based on estimation of corresponding shock pressure together with the assessment of annealing temperatures of pyrometamorphic transformations, is given in Fig. 9. In addition, the rocks were at some elevated temperature at the moment of the impact event due to the geothermal gradient. With this correction, the curve of post-shock temperature should be displaced to the right in Fig. 9 by 120–150 °C, particularly in the lower part.

The temperature regime of annealing is estimated from comparison of the observable mineralogical changes in rocks of the central uplift with experimental data at pressures of $P < 0.1$ GPa that neglected an influence of a steam fluid (Tchukhrov and

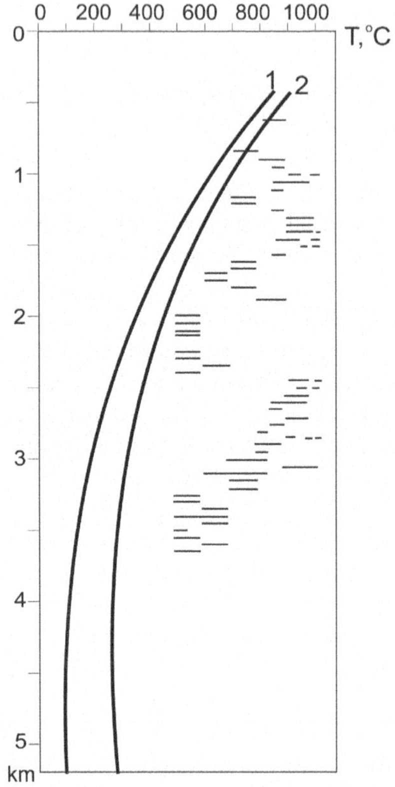

Fig. 9 Estimated temperature of pyrometamorphic transformations through the VDW section. Vertical axis, depth in km; horizontal axis, temperature values. Solid curves show: 1—attenuation of post-shock temperature evaluated base on shock pressure estimates (see Sect. 1); 2—attenuation of the same post-shock temperature with correction to existed geothermal gradient. Horizontal lines show temperature values estimated from petrographic study of core samples (see text)

Bonstedt-Kupletskaya 1965, 1981, 1983; Deer et al. 1978, 1986, 2001). The presence of water as well as the disorder of crystalline structure due to impulse compression would expect a lower temperature of transformations. Therefore, it is reasonable to assume the lower values of the experimentally established temperature intervals.

An appearance of superimposed thermal metamorphism is definitely established in the cases where the observed mineralogical effects cannot be attributed to the residual heat because temperature estimate exceeds considerably an expected post-shock temperature.

In order to clarify the physical parameters of post-shock thermal effects taking into account their imposition on the diaplectic minerals and diaplectic glasses, a series of annealing experiments was performed with biotite-amphibole gneisses naturally shocked up to 45 GPa. The gneisses were sampled from wells drilled within the central uplift. An unshocked gneiss from the Vladimirskoe-2 well situated 70 km northeast of the central uplift (see Fig. 1 in Chap. 2), was used as a reference. Each sample was sawn into 5 pieces, and four of them were heated in a muffle furnace to 500, 750, 950 and 1050 °C and annealed at these temperatures for 30 min with free access of air.

Petrographic study of the naturally shocked to a various degree and subsequently experimentally annealed rocks shows that biotite and amphibole due to iron oxidation first acquire irregular spotty extinction, turn red, become opacitized at the grain edges, and under maximum heating (950 °C), they turn opaque. In gneisses naturally shocked up to 20 GPa, any thermal alteration in quartz and plagioclase were not observed even after annealing at 1050 °C. In rocks compressed up to 25 GPa and higher, planar elements are partially obliterated first in feldspars at 950 °C and then in quartz at 1050 °C. In samples shocked at pressure between 35 and 40 GPa, planar elements in quartz vanish after annealing at 950 °C. The quartz attains spotted extinction and higher birefringence (up to 0.016). The coexisting feldspars are melted in some places after experimental annealing at 1050°. In gneisses shocked at pressures 40–45 GPa, maskelynite is melted at the annealing temperature of 950 °C. In places, it recrystallizes with a fine granoblastic texture. In respective samples annealed at 1050 °C, porous polymineralic glasses with relics of altered plagioclase and quartz appear together with small spots of recrystallized feldspar.

Melanocratic rocks naturally shocked at lower pressure, are more stable during experimental annealing than shocked leucocratic gneisses.

A general scheme of pyrometamorphic changes in diaplectic minerals and glasses and an estimation of respective temperatures are given in Table 2. This scheme was compiled on the basis of petrographic and mineralogical data for thermally transformed rocks of the central uplift. Thermal effects of quartz begin at a temperature of about 573 °C (the inversion temperature of α to β quartz), but they are not reflected in optical properties of newly formed quartz aggregates. Experiments showed that the PDFs disappear when annealed to 600 °C in an atmosphere of steam containing ions of sodium and carbon dioxide (Feldman 1990), but they are stable in dry experimental conditions at 950 °C (Bunch et al. 1968); this was confirmed by our experiments.

Quartz formed by recrystallization of dialectic quartz glass in the rocks from VDW and other wells at temperatures about 600 °C shows a higher birefringence (yellow color interference), and a brown color in transmitted light due to optical dispersion on the boundary between glass and newly formed crystallites. At higher temperature (600–800 °C), spots of micro-granular or globular texture appear in quartz inverted from diaplectic glass (Fig. 10). Metacristobalite, which is a metastable cryptocrystalline modification of β-cristobalite in combination with an amorphous phase, crystallizes from diaplectic quartz and diaplectic glass at a more significant heating. The inversion points for quartz crystal are ca. 1200 °C, and for silica gel, ca. 900 °C (Tchukhrov and Bonstedt-Kupletskaya 1965, 1981). Respective transition temperatures for both diaplectic quartz and, especially, for diaplectic glass are closer to the lower value, but exceed it.

More intensive ($t > 950°$) annealing of quartz and various products of its transformation is established on a basis of zonal distribution of silica modifications in primary

Table 2 Pyrometamorphic transformations of diaplectic minerals and diaplectic glasses

T, °C	500	600	700	800	900	1000	1100
Quartz			Initial recrystallization			Metacristobalite	
Diaplectic glass after quartz			Initial recrystallization			Metacristobalite, Cristobalite	Tridymite
Plagioclase (An_{25-35})		Initial recrystallization		White "ceramics"	Recrystallization		
Diaplectic glass after plagioclase (maskelynite)		Initial recrystallization		White "ceramics"	Recrystallization		
Biotite (f=12-19)				Opacitization rims	Decomposition, recrystallization		
Amphibole (hornblende) (f=33-46)				Opacitization rims	Decomposition, recrystallization		
Garnet (almandine)					Decomposition, recrystallization		

grain boundaries and of tridymite occurrence that is caused not only by high temperature value, but also by rate of its growth (Sinelnikov 1956; Ziegler et al. 1988). A rapid rise of the temperature into the field of tridymite stability (870–1470 °C at 1 atm) results at the beginning in crystallization of metacristobalite rather than tridymite. Then, a stable form of high-temperature cristobalite crystallizes forming needles, prisms, idioblasts, or microspherulites. Finally, at temperature in excess to 1000–1200 °C tridymite with characteristic spear-shaped twins appears in the rocks (Fig. 11). In the center of grains of recrystallized zonal quartz, the transition is completed at the stage of cristobalite formation as a result of the short-term action of the peak temperature. A slower temperature decreasing frequently yields an inversion into quartz with inherited shapes of the replaced phases: lanceolate at the margin zone and an aggregate of isometric rounded individuals in the center of the grain.

The recrystallization of diaplectic plagioclase and diaplectic plagioclase glass (maskelynite) apparently starts at a temperature of about 600 °C, but an appreciable effect of recrystallization of maskelynite is observed only when upon heated to 800 °C (Bunch et al. 1968; Ostertag 1983). At temperatures of 800–950 °C, diaplectic plagioclase and diaplectic plagioclase glass turn into the so-called "white ceramics", which appears brown and dark brown in transmitted light. In places, there are signs

Fig. 10 Thermally-transformed shocked biotite-plagioclase gneiss. In the centre of the image, a partly recrystallized quartz with domains of globular texture and relics of planar deformational elements. The bulk consists of maskelynite (dark-colored) with linearly-orientated isotropized biotite flakes. Photomicrograph, borehole 61, depth 283 m; cross-polarized light

Fig. 11 Lanceolate tridymite twins and granoblastic quartz after diaplectic quartz glass within thermally-recrystallized shocked leucocratic gneiss. *Black*—relics of diaplectic quartz glass. The groundmass consists of granoblastic quartz-feldspar aggregate. Photomicrograph, VDW, depth 1469.5 m; cross-polarized light

Fig. 12 A finely crystalline aggregate of plagioclase microlites after maskelynite. At right bottom and top, tridymite-quartz aggregate after diaplectic quartz glass are observed. The groundmass is partly replaced by saponite. Photomicrograph, VDW, depth 2524 m; cross-polarized light

of melting along one of the twin systems or at grain edges. X-ray diffraction analysis of completely opaque glassy grains shows that they have a crystalline structure with the ordering index of $I \leq 50$ and $I \leq 70$ (samples from the VDW, depths 1575 and 1747 m).

Recrystallization of maskelynite and "white ceramics" results in formation of micro-laths aggregates up to a 0.1 mm across with finely crystalline, in places axiolitic and spherulitic textures (Fig. 12). Maskelynite recrystallization products usually have a more basic composition than the original plagioclase.

The beginning of opacitization of biotite and the change of optical characteristics of hornblende can be attributed to the temperature range of 600–800 °C. Biotite is characterized by narrow opacite rims, while amphibole, by deepening of colour at grain edges. The effect of change in the colour of amphibole is observed during annealing at temperatures of 500–800 °C (Tchukhrov and Bonstedt-Kupletskaya 1983). At higher temperatures, brown and red tints begin to predominate in amphibole and biotite; the higher the level of preceding shock transformation, the more intense the colour appears. The birefringence in hornblende increases to 0.040–0.050. The edges of grains represent optically an opaque cryptocrystalline aggregate, in which monoclinic pyroxene and hematite are recorded by XRD.

A cryptocrystalline opaque aggregate of clinopyroxene and maghemite originate after common hornblende in annealing experiments at temperatures of 1000–1100 °C (Tchukhrov and Bonstedt-Kupletskaya 1983), but for diaplectic amphibole these temperatures may be lower. In the rocks from the VDW, this opaque aggregate consists of clinopyroxene, plagioclase, and iron ore (magnetite, hematite). The clinopyroxene often forms rims of relatively larger microlites around decomposed amphibole

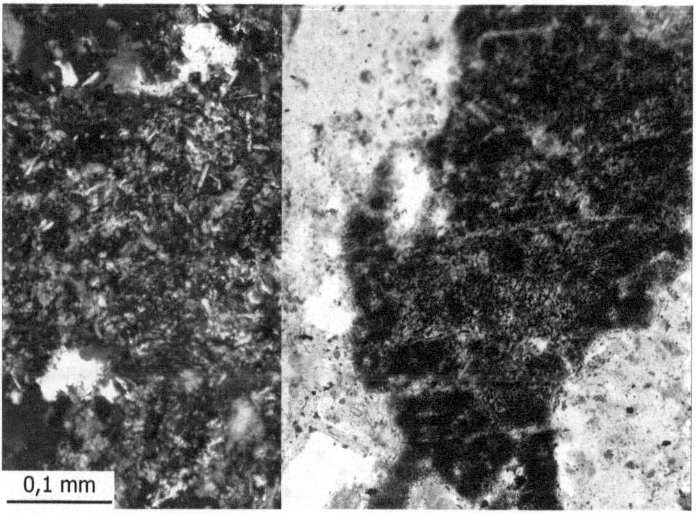

Fig. 13 Opaque aggregate of neogenetic minerals after hornblende (*black*) with rims of fine (<0.01 mm) clinopyroxene (salite) microlites (at right). The opaque aggregate consists of clinopyroxene, plagioclase, magnetite, and hematite. The rock is thermally-recrystallized shocked amphibole gneiss. The texture of the rock is demonstrated at the left side of the image. It consists mainly of feldspar-dominated granoblastic aggregate with quartz-tridymite nests after diaplectic quartz glass. Photomicrograph, VDW, depth 1400 m; cross-polarized (left) and plane-polarized (right) light

(Fig. 13). Their compositions on the polythermal diagram for P = 1 atm (Lindsley 1983) are located close to the isotherm 1000 °C.

Experimental data show that the decomposition of biotite (with an iron content close to biotite from gneisses of VDW) occurs at pressures below 1 kbar at temperatures above 700–800 °C. Biotite is replaced by an aggregate of titanomagnetite, potassium feldspar, more magnesium mica, and occasionally plagioclase. Thus, both input and output of some chemical components takes place during this transformation.

The formation of dusty titanomagnetite and magnetite during the decomposition of femic minerals in annealed rocks resulted in sharp increasing of their magnetic susceptibility, which are recorded by a positive magnetic anomaly (with an amplitude up to 1800 nT) of almost rounded shape and ca. 4 km in diameter in plan. The nature of the magnetic field may indicate that the magnetic disturbing body (in fact, the zone of intensely thermally-metamorphosed rocks) is steeply dipping to SSW; its exposure under the crater fill corresponds in outline to contours of the central pit (Fig. 3). At the crest of the central uplift, natural remanent magnetization values fall sharply, local low-intensity anomalies being recorded at the outer slope of the uplift, though.

The lower stability limit for the almandine garnet at atmospheric pressure is 785 °C (Tchukhrov and Bonstedt-Kupletskaya 1983). In the experiment on thermal stability for a shocked almandine (Gnevushev et al. 1982), a weak exothermic effect caused by oxidation of the ferrous iron was observed at 800–1000 °C. This effect precedes the decomposition of the garnet into a cryptocrystalline aggregate of newly formed minerals including pyroxene and plagioclase.

In individual cases, the high temperatures yield small amounts of melt, which forms at grain boundaries glass films enriched in alkalis. Clinopyroxene similar in composition to that formed at the expense of amphibole, occasionally crystallizes from this melt.

The given data show that assemblages of anhydrous relatively high-temperature minerals are formed as a result of thermal decomposition of hydrous silicates in studied shock-metamorphosed rocks. At the same time, the vapor occurring in the system facilitates easier recrystallization of diaplectic minerals, ensures the transfer of chemical components and yields blasthesis. At final stages, small amounts of newly formed mica and amphibole appear. As it documented by microprobe analyses, an intense re-distribution of alkalis, calcium, magnesium, and, to a lesser extent, silica takes place during this process.

The high degree of post-shock thermal transformations almost completely erases many of the rock features, first of all the texture, and then the mineral composition. The bulk chemical composition remains almost unchanged, and some textural features may be preserved in places. Such thermally transformed shock-metamorphosed rocks, which usually occur in the areas of intensive cataclasis and crushing, have been named c***optoblastoliths*** (from Greek κoπτo—to destroy by blows and βλαστεσισ—shoot) (Masaitis and Mashchak 1996). The appearance of this lithology depends on both the composition of initial rocks and the degree of their preceding shock and thermal (and in places superposed hydrothermal) transformations. Coptoblastoliths are developing mostly at zones where shocked gneisses or amphibolites underwent intense cataclasis and crushing. Due to thermal influence, either massive or porous, fine-grained, dark-colored rocks arisen in these zones; they are eutaxitic frequently (Fig. 14). Coptoblastoliths form irregularly-shaped parts with vague contacts within thermally-transformed shocked rocks. They contribute no more than 5 vol.% to the total length of zones of intense thermal transformation that are recorded mainly in depth intervals of 1000–1100, 1250–1300, 1425–1498, 1830–1910, and 2400–2600 m of the VDW (Masaitis and Mashchak 1996).

The most developed are coptoblastoliths formed at the expense of essentially quartz-feldspathic rocks (Fig. 15); coptoblastoliths after amphibolites and schists as well as after rocks of intermediate composition (mesocratic plagioclase-dominated gneisses), also occur (Fig. 16). All these usually have a spotted appearance due to the presence of relic minerals of original rocks and tridymite porphyroblasts (Fig. 17), which compose about a half of the coptoblastoliths volume. In places where the cataclasis was less intense, relics of gneissic structure can be preserved. The predominant texture of coptoblastoliths is micro-granoblastic combined with micro-ophitic (Fig. 18), ophitic-taxitic, and for rocks formed at the expense of quartz-feldspathic varieties also heteroblastic or felsitic ones (Fig. 15).

The most widespread minerals of coptoblastoliths formed during annealing are clinopyroxene, plagioclase, alkali feldspars, tridymite, titanomagnetite, magnetite, ilmenite, and phlogopite. Clinopyroxene appears as rounded or, less often, short

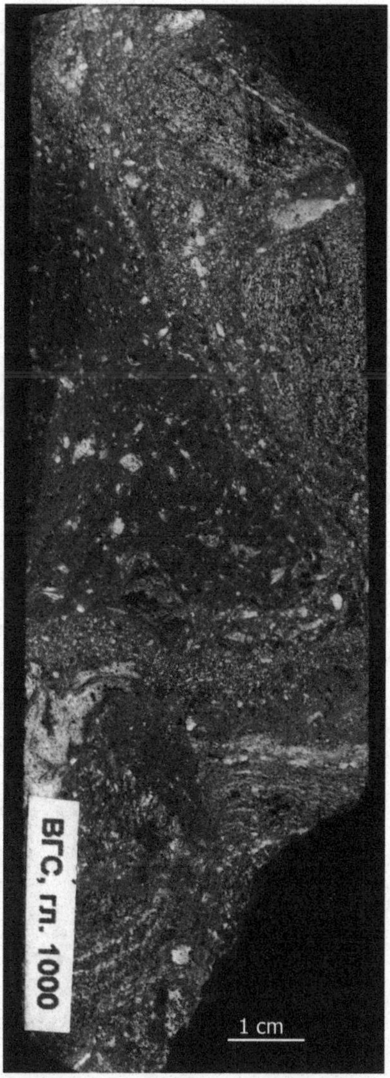

Fig. 14 Intensely thermally-metamorphosed and cataclased biotite-amphibole gneiss changing to completely recrystallized rock (coptoblastolith, dark parts). In the latter, porphyroclasts of recrystallized quartz and feldspar are abundant within a dark fine-grained groundmass. The rock has an eutaxitic structure. VDW, depth of 1000 m

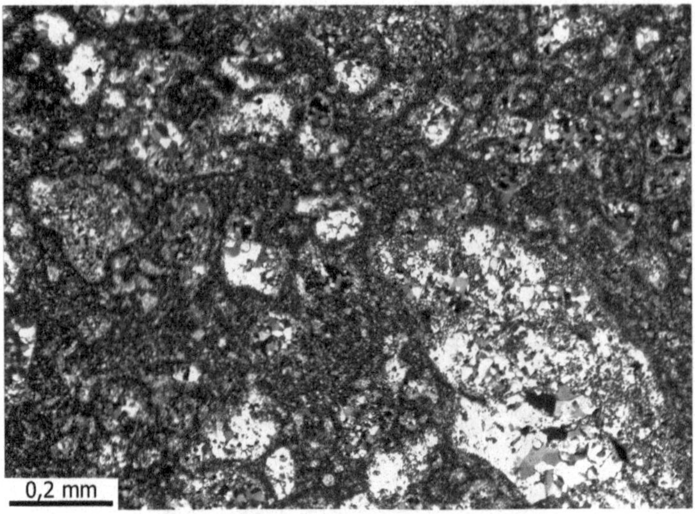

Fig. 15 Coptoblastolith after a leucocratic lithology (granite gneiss). The rock is spotted due to numerous evenly distributed quartz-tridymite glomeroblastic accumulations and rare shadow relict minerals within an inequigranular quartz-feldspar groundmass with heteroblastic texture. The groundmass is partly replaced by saponite, which also fills in small pores. Photomicrograph, VDW, depth 1869 m; cross-polarized light

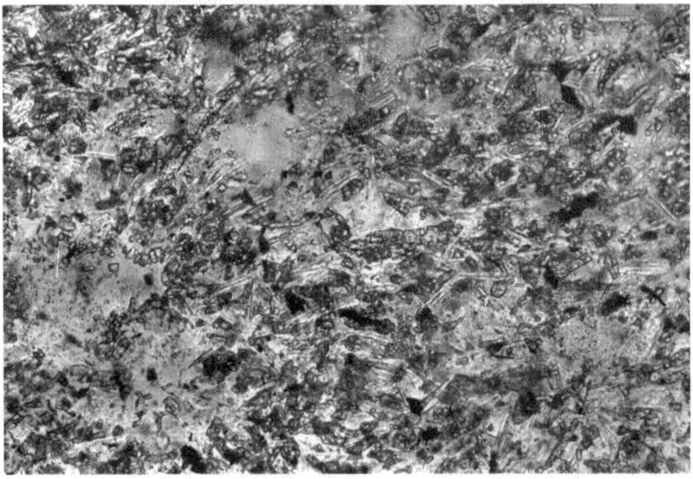

Fig. 16 Coptoblastolith after a mesocratic biotite-amphibole gneiss. Texture is micro-ophitic, formed by plagioclase laths and isometric clinopyroxene microlites. Irregular light spots are granoblastic quartz and feldspar aggregates. Photomicrograph, VDW, depth 1105 m; plane-polarized light

Fig. 17 Coptoblastolith after a migmatized biotite-amphibole gneiss. The rock consists of fine-grained aggregate of feldspars and clinopyroxene microlites hosting tridymite porphyroblasts. Photomicrograph, VDW, depth 1467 m; cross-polarized light

prismatic grains. Its composition it corresponds to salite (Table 3). In this aspect, it differs significantly from pyroxenes from gneisses and amphibolites, as well as from pyroxenes in tagamites (Fig. 18). Plagioclase forms elongated (0.02 × 0.15 mm) sometimes twinned laths of basic oligoclase, or, less often, of acidic andesine composition. Plagioclase in coptoblastoliths after amphibolites is a basic andesine (Table 4). In general, it is more basic than that in the precursor rocks, but it is more acidic than plagioclase in tagamites, which has a higher potassium content (Fig. 19). Phlogopite occurs as short prismatic plates or, less frequently, as flakes. Its composition is even more variable than that of pyroxene and plagioclase, with the Fe/Mg ratio ranging from 1:3 to 1:20 (Table 5).

The mentioned ore minerals demonstrate a spotted distribution and form a dust-like impregnation with particles of 0.001 mm across or isometric segregations of 0.01 mm across, but in places, up to 0.1 mm. Compositions of magnetite and titanomagnetite are given in Table 5.

Alkali feldspars fill the interstices between plagioclase laths and they are also preserved as recrystallized relicts of primary minerals. These relicts are characterized by broom-like and spherulite textures. Feldspars, which fill interstices have microgranoblastic, radial, lath-like and other textures. The composition of K-feldspars is variable, orthoclase and anorthoclase are present (Table 4). Tridymite occurs as spear-shaped twins within the spotted areas of granoblastic quartz; its presence is particularly typical for leucocratic coptoblastoliths (Fig. 17).

The chemical composition of coptoblastoliths is generally similar to that of the precursor crystalline rocks. The data points for coptoblastoliths plot in the AFM ternary diagram are located in the same field as the points of shocked crystalline

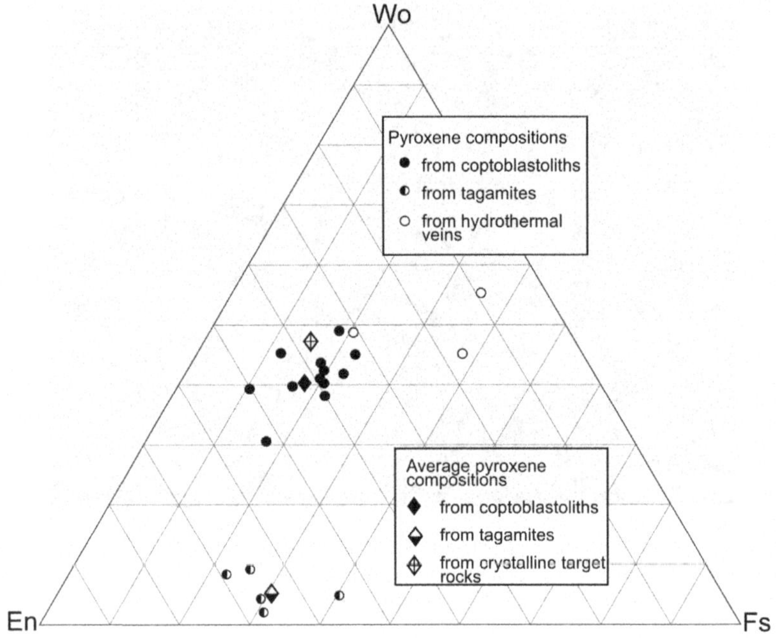

Fig. 18 Ternary Wo-Fs-En diagram for clinopyroxenes from rocks of the central uplift of the Puchezh-Katunki impact structure. Dots for pyroxenes from coptoblastoliths, tagamites, post-impact hydrothermal associations as well as dots of mean composition for pyroxenes from target crystalline rocks (*see* Masaitis and Pevzner 1999), tagamites and coptoblastoliths are plotted. EMP analyses of pyroxenes are given in Tables 5 (in Chap. 3), 3, and 8

rocks that host the zone of coptoblastoliths (see Fig. 36 in Chap. 3). As noted above, coptoblastoliths show signs of initial selective melting. This melt is relatively more acid and alkaline, and enriched in volatiles. Thin veinlets of such granophyre (up to 2–3 cm thick) with sharp contacts are found in coptoblastoliths at depths of 1380 and 1460 m in the VDW.

These granophyres have spotted and fluidal texture, they contain irregular and lenticular voids filled with heulandite and chalcedony. The groundmass is fine-crystalline, prismatic-grained, in some places spherulitic, and composed of pyroxene ($Fs_{19}En_{40}Wo_{41}$), alkali feldspar and tridymite. Compared to host shock- and thermally-metamorphosed gneisses, granophyres are distinguished by higher contents of silica and alkalis, and lower contents of femic components (Table 6). The granophyres may be considered as rheomorphic injections.

The pressure–temperature (PT) conditions of the coptoblastoliths formation are similar to those of the pyroxene-hornfels facies of contact metamorphism for metapelite, quartz-feldspathic and basic rocks. Maximum temperatures of metamorphism of this facies at low pressures are up to 1100 °C. Rocks similar to those found in the Puchezh-Katunki have been found in central uplifts of the Boltysh and Terny impact structures in the Ukraine (Masaitis et al. 1980; Masaitis and Mashchak 1994),

Table 3 Selected electron microprobe analyses of clinopyroxenes from coptoblastoliths

Components	Sample numbers										
	1336	1399	1470	1481	1486.4	2470	2472	2502	2568	2578	2596
SiO_2	50.14	51.74	51.09	55.63	52.37	52.10	52.89	49.69	52.90	51.70	51.67
TiO_2	0.52	0.33	0.48	0.15	0.12	0.26	0.44	0.44	0.27	–	0.32
Al_2O_3	6.33	0.46	0.58	1.19	1.51	1.06	–	1.96	0.35	0.73	0.20
FeO	10.46	11.52	11.97	6.75	7.90	10.91	13.21	11.70	13.61	14.14	13.89
MnO	0.21	0.45	0.42	0.32	0.39	0.08	0.24	0.25	–	0.18	0.36
MgO	17.67	13.64	14.00	17.12	14.84	15.11	13.57	11.02	13.69	10.75	12.40
CaO	13.96	20.94	20.38	18.51	21.84	19.36	19.33	23.57	18.77	22.39	20.56
Na_2O	0.32	0.20	–	0.48	0.69	0.73	0.17	0.76	0.38	0.15	0.21
K_2O	0.24	–	–	–	–	–	–	–	–	–	–
Total	99.85	99.28	98.92	100.15	99.66	99.61	99.85	99.39	99.99	100.04	99.61
Fs	17	18	19	11	13	17	21	19	22	23	22
En	53	39	40	50	42	43	39	32	39	31	36
Wo	30	43	41	39	45	40	40	49	39	46	42
n	3	4	3	3	5	4	3	3	1	3	5

Electron microprobe data, analyst L.S. Bogomolnaya (VSEGEI)
The sample number is the depth in VDW section. n is number of repeated analyses

Table 4 Selected electron microprobe analyses of feldspars from coptoblastoliths

Components	1	2	3	4	5	6	7	8	9	10	11	12	13	14	15
	Sample numbers														
	62/185	1399	1470	2470	2472	2568	1399	1399	1399	1465	1465	1481	1481	2470	2568
SiO_2	61.88	67.01	57.37	62.10	64.53	59.29	64.09	65.26	66.95	62.76	62.35	64.90	66.63	65.46	64.10
TiO_2	–	0.25	–	0.10	–	–	0.50	0.25	0.52	–	–	–	–	–	0.43
Al_2O_3	22.33	19.55	25.58	23.00	20.98	26.46	20.62	17.69	16.85	21.70	21.50	20.18	19.27	18.78	18.64
FeO	–	0.71	–	0.63	0.32	0.43	0.19	0.74	0.99	0.84	0.92	–	–	–	–
MnO	–	–	0.67	–	–	–	–	–	–	–	–	–	–	–	–
CaO	5.05	7.00	8.26	4.81	5.35	9.73	1.90	–	0.37	3.39	3.72	0.76	0.80	0.55	0.38
Na_2O	9.31	3.27	7.07	8.31	7.91	4.54	8.01	1.17	0.58	9.02	8.27	6.23	5.90	5.54	–
K_2O	1.07	1.73	0.57	0.98	0.36	0.56	4.20	14.95	14.46	1.95	1.72	7.67	7.57	9.34	16.16
Total	99.64	99.52	99.52	99.93	99.45	101.01	99.51	100.07	100.78	99.66	98.48	99.75	100.18	99.66	100.01
An	22	47	38	23	27	53	9	0	2	15	18	4	4	3	2
Ab	73	39	59	72	71	44	67	10	5	74	72	53	52	46	0
Ort	5	14	3	5	2	3	24	90	93	11	10	43	44	51	98
n	3	3	3	2	1	1									

Samples 1–6 are plagioclases; 7, 10, 11—albite (with orthoclase component); 8, 9, 15—orthoclase, 12–14—K-Na-feldspars
Electron microprobe data, analyst L.S. Bogomolnaya (VSEGEI)
For samples 2–15, the number is the depth in VDW section. The sample 1—from bh. 62, depth 185 m. *n*, number of repeated analyses

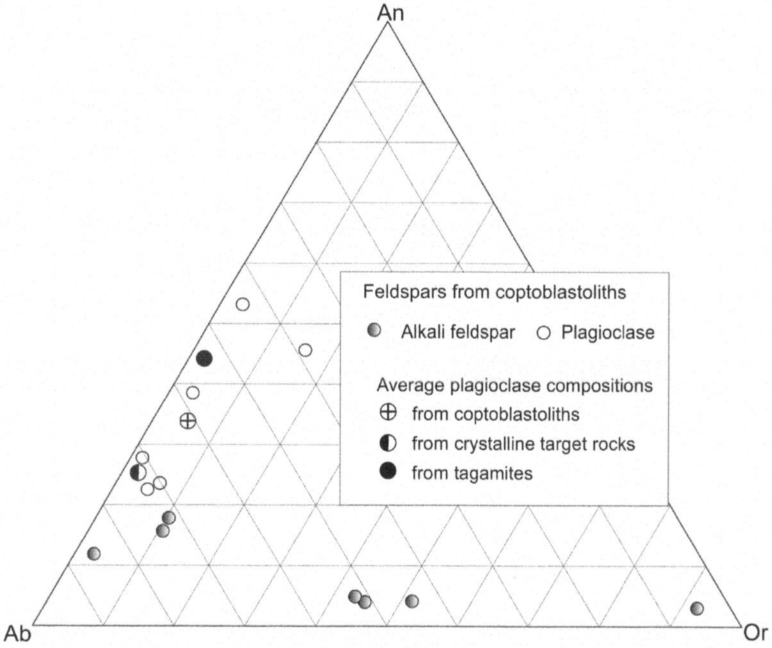

Fig. 19 Ternary An-Ab-Ort diagram for feldspars from rocks of the central uplift of the Puchezh-Katunki impact structure. In addition to feldspars from coptoblastoliths, mean values for plagioclases from crystalline target rocks (*see* Masaitis and Pevzner 1999), coptoblastoliths, and tagamites are shown. EMP analyses of feldspars are given in Tables 5 (in Chap. 3) and 4

they also may be considered as coptoblastoliths (Masaitis and Mashchak 1994). Peculiar annealed crystalline rocks (so-called leucogranofelses) in the central uplift of the Vredefort impact structure (Schreier 1983; Hart et al. 1995; Gibson et al. 1998, and others) also resemble the above described coptoblastoliths by many features. The annealing temperatures of these rocks are estimated at 900–1000 °C, which are also significantly higher than the residual post-shock temperatures there. In some cases, the annealing resulted in partial melting and remobilization of a relatively leucocratic melt there.

The coptoblastoliths of the deep sub-crater zones of the Puchezh-Katunki and some other impact structures may be correlated with the so-called lunar granulites in a number of structural and textural features and reconstructed conditions of transformation of initial rocks (Masaitis and Mashchak 1995). The lunar granulites are recrystallized polymict or monomict impact breccias with a granoblastic texture formed after ferroan anorthosites, norites and troctolites. Their origin is disputed (e.g., Lindström and Lindström 1986; McGee 1987, 1989; James et al. 1989, and others). The most likely that the formation of coptoblastoliths as well as lunar granulites is stimulated by heating caused by intense shifting stresses accompanied by differential movements under conditions of unloading after pressure decay. Differences between the two types of recrystallized impact breccias—terrestrial and

Table 5 Selected electron microprobe analyses of phlogopite and ore minerals from coptoblastoliths

Components	1	2	3	4	1	2	3	4	5
	Sample numbers								
	1336	1481	1484	1486.4	1470	1470	1470	1481	1481
SiO_2	37.07	42.78	42.67	41.94	–	–	–	1.28	0.99
TiO_2	2.58	2.77	1.93	3.97	0.69	0.83	9.81	8.20	9.29
Al_2O_3	14.48	11.87	12.99	14.45	–	–	–	1.59	1.90
FeO	12.97	5.75	2.72	8.05	98.75	99.14	89.90	78.62	79.60
MnO	–	–	–	–	–	–	–	0.88	–
MgO	20.28	23.71	26.46	18.54	–	–	–	1.96	2.31
CaO	1.53	0.14	0.22	0.24	–	–	–	0.27	–
Na_2O	–	0.45	2.04	1.59	–	–	–	1.69	0.85
K_2O	7.26	8.38	6.78	8.86	–	–	–	–	–
Sum total	96.17	95.85	95.81	97.64	99.44	100.19	100.13	94.49	94.94
Fe	27	12	5	20					
Mg	73	88	95	80					
Fe/Mg	2.7	7.3	19	4					
n	2	2	2	2	1	1	1	1	1

1–4—phlogopite, 5 and 6—magnetite, 7–9—titanomagnetite
Electron microprobe data, analyst L.S. Bogomolnaya (VSEGEI)
The sample number is the depth in VDW section. *n* is number of repeated analyses

Table 6 Whole rock analyses (wt%) of a granophyre and host coptoblastolith after biotite-amphibole gneiss (VDW, depth 1380 m)

Components	Granophyre	Coptoblastolith
SiO_2	64.28	59.02
TiO_2	0.38	0.53
Al_2O_3	14.77	13.8
Fe_2O_3	1.61	3.18
FeO	0.72	2.17
MnO	0.04	0.11
MgO	1.79	5.07
CaO	4.57	9.07
Na_2O	3.89	4.48
K_2O	4.02	0.76
P_2O_5	–	0.11
L.O.I.	4.32	2.11
Total	100.39	100.41

lunar—result mainly from differences in the composition of their precursor rocks, in their fluid content, and in the repeated character of impacts on the Moon. The inferred original geological position of lunar granulite samples may have been the same as that of their terrestrial analogues.

One more specific appearance of thermal transformation of shocked crystalline rocks in the Puchezh-Katunki should be noted. This is a near-surface oxidation, which took place prior to the formation of hydrothermal veins. The oxidation is appearing in the annular crest of the central uplift, i.e. in the most elevated part of the latter. From drilling data, the depth of the zone of oxidation reaches 200 m. It is manifested by ochre colour of shocked and brecciated crystalline rocks due to formation of hematite and limonite after mafic minerals and magnetite. It is quite possible that hematite and limonite developed in the most elevated part of the authigenic breccia as a result of interaction of the rocks heated up to 200–400 °C and atmospheric oxygen immediately following the origin of the central uplift, inasmuch as this part of the uplift was elevated to the surface.

3 Hydrothermal Alteration

Evidence of impact-induced hydrothermal activity includes both occurrence of secondary mineral assemblages (including ore-forming minerals) and chemical alteration of materials at the impact site. Due to drilling, secondary mineralization can be traced in the Vorotilovo Deep Well to 5.3 km depth, as well as laterally in the area of the central uplift. The degree of hydrothermal alteration of rocks depends on both their permeability and degree of preceding thermal recrystallization; Hydrothermal minerals commonly contribute no more than 1–2% to the rock volume but in places, they peak 30 vol.% in coptoblastoliths. Hydrothermal alterations become less pronounced downwards and outwards reflecting the rather regular decrease in rock porosity. Thus, noticeable appearance of hydrothermal alteration minerals does not extend beyond the central uplift of the Puchezh-Katunki astrobleme.

The Puchezh-Katunki impact structure is stand out for its diversity of hydrothermal minerals. The list of the latter includes smectites, chlorite, illite, calcite, sulfides, zeolites, anhydrite, gypsum, apophyllite, quartz, opal, prehnite, epidote, andradite, ferrosalite, actinolite, and albite. The spatial distribution of the hydrothermal mineralization is characterized by both vertical and lateral zonation. The vertical zoning is most completely represented in the middle part of the central uplift, where it was studied in the VDW section. There, two alteration zones, named after the main minerals that occur in these zones, can be distinguished: the upper smectite-zeolite zone and the lower chlorite-anhydrite zone (Fig. 20).

The smectite-zeolite zone comprises layers of coptomict deposits, suevites, and polymict allogenic breccia, plus the shocked and brecciated target rocks to 2550–2600 m depth. The compositions of main minerals of the zone—both smectites and, especially, zeolites vary regularly with depth (Table 7).

According to DTA, XRD, and microprobe data, montmorillonite, Mg–Fe-saponite, and nontronite are distinguished among the smectites (Naumov 2002). Montmorillonite (b_0 = 8.98 Å) replaces impact glass fragments in suevites and redeposited suevites, and fills locally interstices in the authigenic breccia and in tagamites in peripheral parts of the central uplift. Mg–Fe-saponite (b_0 = 9.18–9.22 Å) replaces ubiquitously shocked and thermally transformed rock-forming minerals in the authigenic breccia and is developed after the matrix and in tagamite interstices. With depth, the Al^{IV} content increases in the saponites. Nontronite in association with calcite occurs as thin veinlets cutting saponitized rocks. For smectites, this variation appears in the progressive substitution of dioctahedral high-alu-mina varieties by trioctahedral Mg–Fe varieties (saponites) downward in the section, with the content of tetrahedral alumina increasing with depth in the saponites. Thus, a general compositional variation for smectites appears in substitution of dioctahedral Al-montmorillonite by trioctahedral Mg–Fe varieties (saponites) downward in the section, the content of tetrahedral aluminia increasing with depth in these saponites. Finally, at the depths of 2500–2600 m saponites are gradually replaced by chlorites (Fig. 20). So, the boundary between the smectite-zeolite, and the chlorite-anhydrite zone is drawn at this level. XRD data failed to reveal the presence of chlorite-smectite (mixed-layered phases in this interval.

The zeolites form numerous veins up to 10 cm thick in the authigenic and polymict allogenic breccias and suevites and locally, form the cement of coptomict gritstones; they associate with three generations of calcite, and with pyrite, apophyllite, and gypsum. Among zeolites, heulandite, chabazite, analcime, stilbite, laumontite, clinoptilolite, erionite, phillipsite, harmotome, and scolecite are identified by EMP, XRD, and DTA analyses (Naumov 2001).

Both a vertical zonation and some lateral inhomogeneity of the distribution of zeolites is noted. On the whole, the Ca/Na and Al/Si ratios in zeolites increase downwards in the section (Table 7, Fig. 21), so that some zeolite subzones can be

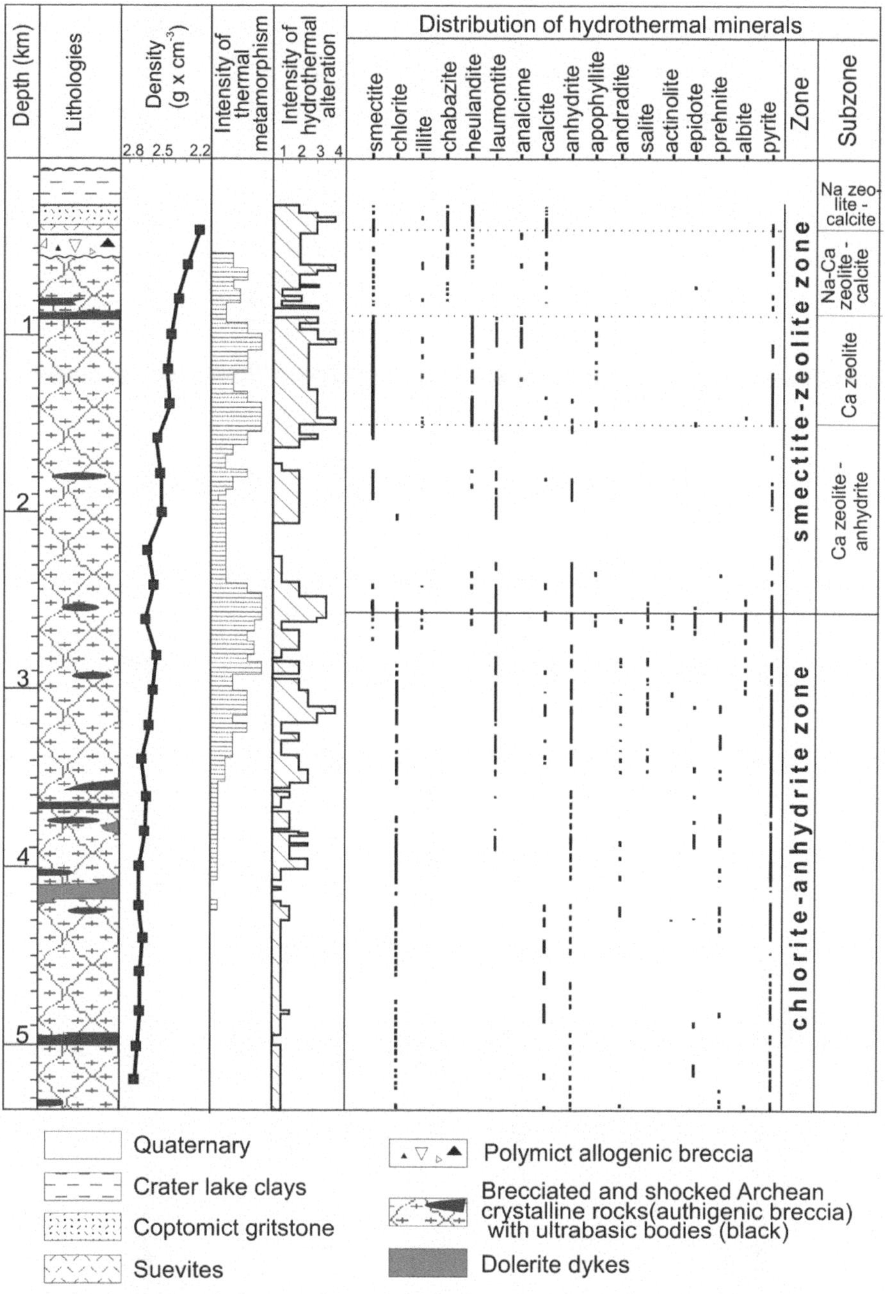

Fig. 20 Distribution of main hydrothermal minerals in the Vorotilovo Deep Well section. The intensity of thermal metamorphism is shown based on data from Sect. 2 in arbitrary units by rectangle areas for 50-m depth intervals. The intensity of hydrothermal alteration is in arbitrary units, reflecting the content of secondary minerals calculated for 15-m depth intervals from thin section study: 1, <0.5%; 2, 0.5–2%; 3, 2–4%; 4, >4%

Table 7 Representative EMP analyses of main minerals from the smectite-zeolite zone

Subzone	Na-zeolites-calcite				Na–Ca-zeolites-calcite								Ca-zeolites		
Mineral	Montmor illonite	Calcite	Heulandite	Chabasite	Saponite	Calcite 1	Calcite 2	Calcite 3	Heulandite	Chabasite	Analcime	Apophyllite	Saponite	Laumontite	Heulandite
No.	1	2	3	4	5	6	7	8	9	10	11	12	13	14	15
Borehole	44	99	VDW	VDW	55	754	63	754	54	61	33	C3	VDW	VDW	VDW
Depth (m)	275	99/365	337	334	244	188	271	209	297	295	421	559	1404	1467	1408
SiO_2	61.04	–	60.06	53.87	39.57	–	–	–	59.07	48.42	57.13	53.01	40.34	52.26	55.15
TiO_2	0.31	–	0.03	–	–	–	–	–	–	–	0.02	0.02	–	–	–
Al_2O_3	14.56	–	14.86	14.23	6.31	–	–	–	15.39	18.68	23.08	0.73	7.58	21.69	17.25
FeO[a]	8.18	0.36	–	0.10	22.74	0.11	–	2.12	0.12	–	0.25	0.23	23.50	–	–
MnO	0.28	0.40	–	–	0.29	0.06	0.63	0.60	–	–	–	–	–	–	–
MgO	6.69	0.59	–	0.05	11.98	0.01	0.25	0.08	0.26	–	–	–	12.94	–	–
CaO	2.35	53.09	3.43	4.32	1.96	54.82	54.77	53.35	4.90	6.26	–	24.98	3.40	11.82	6.94
Na_2O	0.82	–	2.42	8.55	0.53	–	–	–	3.11	3.64	11.86	0.28	–	0.52	1.66
K_2O	0.21	–	0.73	0.75	0.06	–	–	–	0.55	0.37	0.01	4.83	–	0.16	0.21
BaO	N.d.	0.25	N.d.	N.d.	N.d.	–	–	–	0.19	–	–	–	N.d.	–	0.70
SrO	N.d.	–	N.d..	N.d.	N.d.	–	–	–	1.65	0.33	0.01	0.02	N.d.	0.08	1.97
H_2O	N.d.		–	N.d.	N.d.	–	–	–	14.82	21.93	6.88	13.99	N.d.	12.78	16.26
CO_2	N.d.	45.20	N.d.	N.d.	N.d.	43.20	44.36	44.85	N.d.	N.d.	N.d.	N.d.	N.d.	N.d.	N.d.
Total		99.46				98.20	100.01	101.05							

[a]Total iron as FeO

Selected representative analyses are shown. Analyses 2, 7, 8, 9, 10, 11, 12,15 are performed by XRF with SRM-2 instrument, analysts are B.A. Tsimoschenko, V.N. Tarasova, A.B. Safro, and I.G. Ginzburg, VSEGEI Lab. Other analyses, EMP data, analyst L.S. Bogomolnaya, VSEGEI Lab

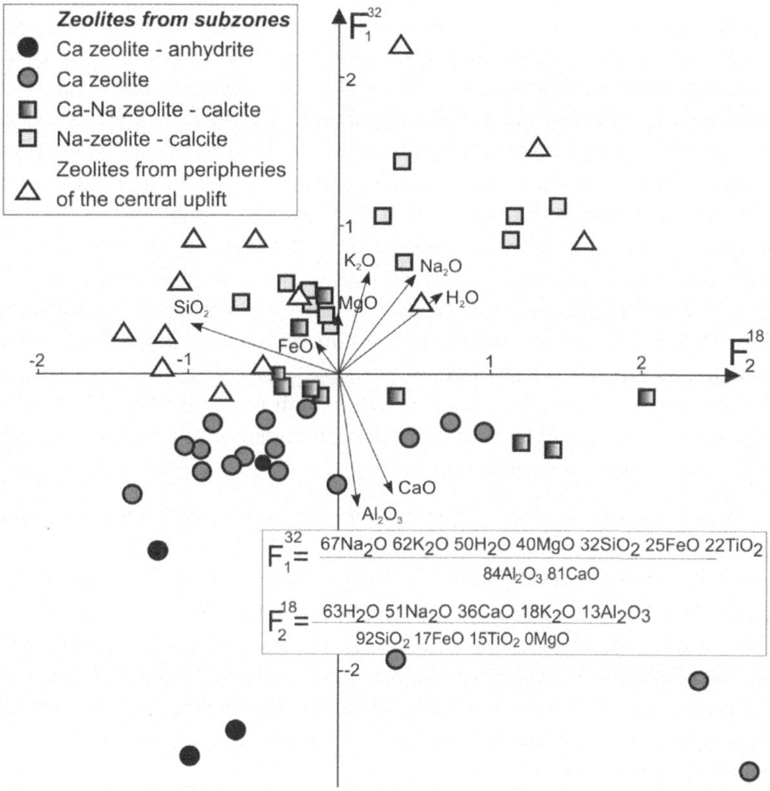

Fig. 21 Diagram of main components for chemical compositions of zeolites from different parts of a generalized vertical section through the central uplift of the Puchezh-Katunki impact crater. Results of 58 X-ray fluorescence spectroscopy and electron microprobe analyses are used. Representative chemical data of zeolites are given elsewhere (Naumov 1993, 2002). The data show a general increase of Ca and Al contents in zeolite compositions downward in the vertical section as well as the rating of zeolites from peripheral segments of the central uplift to Na zeolite–calcite subzone

distinguished, from top to bottom (Fig. 20):

(a) *The Na zeolites-calcite subzone* embraces redeposited suevites, and, locally, underlying suevite layer. Zeolites are represented by Na-heulandite, Na-chabazite, and, in some locations, erionite and clinoptilolite. Together with calcite, they form the cement of the redeposited suevites and suevites within the central pit and the disseminated mineralization in these rocks within the annular depression. Among clay minerals, which replace impact glass, montmorillonite dominates over saponite.

(b) *The Na–Ca zeolites—calcite subzone* contains suevites, polymict allogenic breccia, and shocked and brecciated basement rocks to 550–850 m depth. In this subzone, zeolite-bearing assemblages vary laterally, that indicates a significant chemical variability of the mineral-forming solutions. Common are heulandite-calcite, analcime-calcite, chabazite-analcime-(heulandite)-calcite, stilbite-analcime, apophyllite-analcime-heulandite-gypsum, apophyllite-erionite-heulandite associations. Just within this subzone, the radial variation of the zeolite distribution is revealed (Naumov 2002). It consists in both an alternation of areas with high-silica zeolites (heulandite, erionite) and low-silica zeolites (chabazite, analcime), and in increase of the Si and Na contents in chabazite, a zeolite occupying only a thin layer (no more 100 m thick) at the top of the authigenic breccia, outward from the crater center (Naumov 1993). Clay minerals are mainly represented by saponite within this subzone.

(c) *The Ca zeolites subzone* is penetrated by only VDW core in the depth interval 850–1800 m. Laumontite and Ca-heulandite, with minor apophyllite, chabazite, scolecite, and quartz occur there.

(d) *The Ca zeolites—anhydrite subzone* (the depth interval 1800–2550 m) is distinguished by the appearance of anhydrite, together with laumontite and pyrite, as a common secondary mineral.

The chlorite-anhydrite zone comprises brecciated and shocked basement rocks at depths more than 2550 m. Common are chlorite, anhydrite, and pyrite, and locally (mainly in the depth interval 2900–3500 m), relatively high-temperature Ca–Fe-silicates (andradite, ferrosalite, epidote, prehnite, actinolite), and, in some cases, quartz also occur. The representative chemical analyses of main minerals from this zone are given in Table 8. Zeolites are observed down to 3854 m and are represented by Ca-varieties (laumontite with minor heulandite) only in this zone. Below 4200 m, the hydrothermal mineralization is rarely to be found and represented by thin (no more than 1–2 mm thick) veinlets.

Along the periphery of the central uplift, the vertical zoning is much less appeared. In places, a superposition of characteristic minerals from different zones (e.g., zeolites, smectites and epidote, actinolite) is recorded there. In addition, a reverse zoning is observed in some boreholes where chlorites replaces smectites downward in the section, whereas zoning of zeolite distribution fits in general into the patterns described above (Fig. 22).

Among post-impact alteration minerals, sulfides are paid attention due to its economic significance in some impact structures (Johansson 1984; Masaitis 1989; Grieve and Masaitis 1994). Within shocked basement rocks, some primary (pre-impact) sulfides occur; these are mostly pyrite and subordinate pyrrhotite, but in rare places, sphalerite, chalcopyrite, pentlandite, and millerite are documented. Post-impact sulfides are represented by three assemblages: (a) chalcopyrite-sphalerite-galena-pyrite association in laumontite-anhydrite veins that are common at the depth from 1400 to 3200 m in the VDW section; (b) marcasite-pyrite at marginal parts of analcime-calcite veins occurring at the top of authigenic breccia and overlying suevite and lithic allogenic breccia; (c) pyrite with minor marcasite and very rare pyrrhotite within the matrix of suevite and coptomict gritstone. Thus, pyrite is a throughout sulfide for all hydrothermal associations. Correspondingly, three pyrite generations occur distinguished by localization, crystal habit, chemical composition and sulfur isotope ratio.

Table 8 Representative EMP analyses of main minerals from the chlorite-anhydrite zone (VDW section)

Mineral	Chlorite	Andradite	Ferrosalite	Epidote	Actinolite	Prehnite	Calcite	Albite	Laumontite	Heulandite
Sample numbers	2578	3101	2806	3086	3437	3437	1412	1412	2410	2806
SiO_2	29.17	35.74	52.16	38.34	55.19	45.11	–	63.64	56.73	58.59
TiO_2	–	–	–	0.46	0.15	–	–	–	–	–
Al_2O_3	17.19	0.12	2.37	24.44	2.03	24.24	–	24.19	23.86	17.10
Fe_2O_3	–	30.04	–	9.95	5.26	–	–	–	–	–
FeO	24.83	–	19.23	2.15	7.88	0.68	0.23	–	–	–
MnO	0.24	0.10	0.58	0.71	0.40	–	0.72	–	–	–
MgO	17.74	0.07	2.56	–	15.62	0.15	0.16	–	0.23	0.21
CaO	0.29	33.39	22.79	22.08	11.22	24.80	54.90	0.06	11.99	6.44
Na_2O	0.39	0.13	0.85	–	–	0.58	–	11.16	–	0.54
K_2O	–	0.16	–	–	–	0.15	–	–	–	0.44
Total	90.11	99.75	100.54	98.13	97.75	96.01	56.01	99.04	92.58	83.22
n	3	5	2	2	2	2	1	2	2	4

EMP analyses, analyst L.S. Bogomolnaya, VSEGEI Lab
Numbers of samples correspond to the depth in the VDW section. *n* is number of repeated analyses

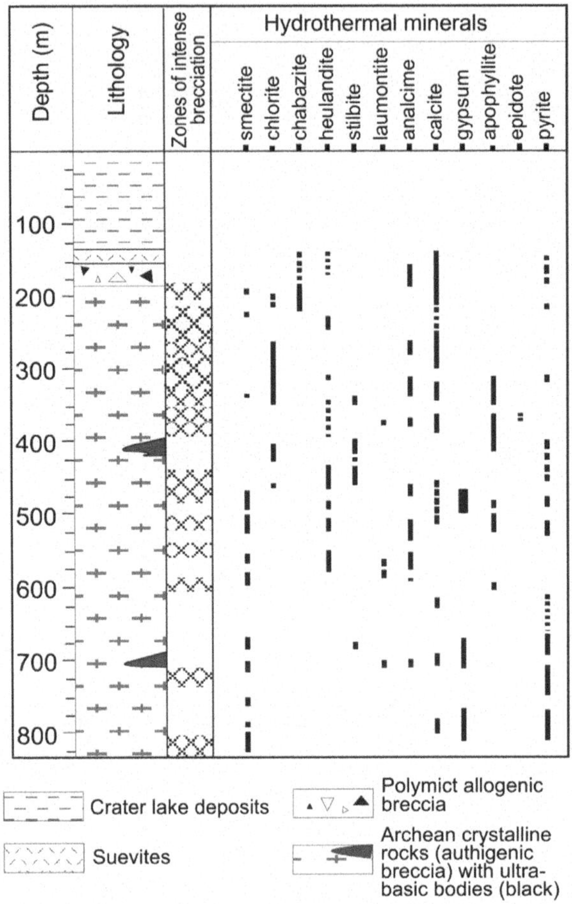

Fig. 22 Distribution of hydrothermal minerals in a vertical section of the southern part of the central uplift of the Puchezh-Katunki impact structure (drill cores 40, 63, 754). Dashed lines indicate sporadic occurrence of a mineral. Compared with the Vorotilovo Deep Well (VDW), the inverse distribution of clay minerals is established. The distribution of zeolites and associated minerals is similar to the VDW section

(1) Disseminated and vein pyrite from authigenic breccia (down of 1500 m in the VDW section) is characterized by: (a) prismatic and tabular crystal habit; (b) yellow color of grains; (c) high (>1) Co/Ni ratio; (d) $\delta^{34}S$ values (from −0.5 to +3.3‰ CDT) are typical for basic mantle-derived igneous rocks (Seal 2006) to inherit more probably the sulfur isotope composition of pre-impact sulfides from the Precambrian basement rocks (Table 9).

(2) Pyrite from veins cutting allogenic breccia, suevite, and intensely deformed and shocked target rocks at the top of the central uplift, is represented by aggregates of fine (<0.3 mm) rhombidodecahedrons and cuboctahedrons of bronze tint;

Table 9 Isotopic compositions of sulfur, oxygen, and carbon in some hydrothermal alteration minerals from the Puchezh-Katunki impact structure

No.	Bore-hole.	Depth (m)	Mineral	Sample description	$\delta^{13}C$, ‰ (V-PDB)	$\delta^{18}O$, ‰ (SMOW)	$\delta^{34}S$, ‰ CDT
1	40	257	Calcite	Calcite-chabasite veins in brecciated gneiss	−22.66	22.61	
2	63	315	Calcite		−8.70	21.78	
3	40	300	Calcite		−20.30	21.90	
4	754	216	Calcite	Yellow rhombohedral calcite from veins in cataclased gneiss	−2.96	24.94	
5	63	271	Calcite	Pyrte-calcite-analcime vein in brecciated gneiss	−55.40	22.40	
6	754	209	Calcite	White calcite—a vein in brecciated gneiss	−1.00	18.56	
7	754	188	Calcite		−3.80	24.60	
8	54	160	Calcite	Yellow calcite geodes in brecciated gneiss	−6.45	24.50	
9	60	245	Calcite	Scalenohedric calcite in vugs of allogenic breccia cement matter	−6.72	18.00	
10	65	161	Calcite		−14.61	25.76	
11	99	365	Calcite	Yellow calcite cementing the matrix of coptomictic gravelite	−11.59	23.03	
12	38	478	Pyrite	Dissemination in groundmass of coptomict gritstone			33.4
13	50	314	Pyrite	Pyrite-quartz-calcite vein cutting allogenic breccia			−8.8
14	14	260	Pyrite	Sulfide dissemination in cataclasite			−15.4

(continued)

Table 9 (continued)

No.	Bore-hole.	Depth (m)	Mineral	Sample description	$\delta^{13}C$, ‰ (V-PDB)	$\delta^{18}O$, ‰ (SMOW)	$\delta^{34}S$, ‰ CDT
15	62	264	Pyrite	Disseminated in intensely shock- and thermally-metamorphosed gneiss			−8.6
16	33	421	Pyrite	Pyrite-analcime-calcite vein cutting shocked gneiss			−21.6
17	754	403	Pyrrhotite	Sulfide veinlets and dissemination in cataclased amphibolite			−0.5
18	VDW	3524-1	Pyrite	Sulfide dissemination in shocked biotite-amphibole gneiss			3.2
19	VDW	3524-2	Pyrite				3.2
20	VDW	3524-3	Pyrrhotite				3.3
21	VDW	2567	Pyrite	Dissemination and veinlets in shocked amphibolite			2.2
22	VDW	2850	Pyrite	Pyrite-laumonite-anhydrite vein			2.4
23	VDW	2850	Anhydrite			10.4	12.9
24	VDW	3100	Pyrite	Pyrite-laumonite-anhydrite vein			2.8
25	VDW	3100	Anhydrite			7.9	11.1
26	VDW	3854	Pyrite	Dissemination in brecciated gneiss			2.7
27	VDW	3854	Anhydrite			7.0	10.7
28	VDW	4252	Pyrite	Dissemination in brecciated gneiss			2.4

For samples 3, 5, 7, 21–28, isotopic compositions were determined by R.E. Prilutsky (VSEGEI Lab.). The isotope composition of carbon and oxygen in calcite (samples 1, 2, 4, 6–11) is measured by E. Versh, University of Tartu, Estonia (Versh et al. 2006, and pers. comm.). The sulfur isotope composition in sulfides (samples 12–20) is measured by V.A. Grinenko (GEOHI RAS, Moscow)

Co/Ni ratio is low (<1). This pyrite is enriched by the light sulfur isotope ($\delta^{34}S$ varies from −8.6 to −21.6‰ CDT; Table 9).

(3) Pyrite in basal crater lake sediments (coptomict gritstone) is characterized by cubic or, more rarely, by pentagon-dodecahedron habit and of no more than 0.1 mm in size; in addition, anhedral bronze-tinted grains occur. Pyrite is highly enriched by the heavy sulfur ($\delta^{34}S = 33.4$‰ CDT) (Table 9).

Both vertical zonation of post-impact sulfide associations and high variability in the composition and crystal habits indicate to the significant variation of crystallization conditions within the hydrothermal cell while distinction of sulfur isotope composition is evidence of different sources of fluids.

The estimation of input and output of chemical components during the hydrothermal transformation showed that chemical alteration trends are very similar for lower (chlorite-anhydrite) and upper (smectite-zeolite) zones of the hydrothermal column (Masaitis and Pevzner 1999): an addition of Ca and Mg and depletion of alkali and silica are observed for both ones. The principal differences between the zones are the less important depletion of silica in the smectite-zeolite zone, and the inversely correlated behavior of Al and Fe there. Aluminium is added in the smectite-zeolite zone, but iron, in contrast, in the chlorite-anhydrite zone. The summary effect includes significant addition of Ca, Mg, Sr, La, and Y, minor addition of siderophile elements (with increasing intensity in the Co-Cr-Ni series), and also Mo, Cu, Zn, and V. This is correlated with a loss of K, Na, Si, Ti, P, Al, Mn, Ga, Pb, and W, whereas, some of the lithophile elements (Nb, Zr, Yb, and Sc) are inert. Iron increases in the lower zone of the hydrothermal column together with other siderophile elements, but it is depleted in the upper zone. The balance between gain and loss of components is consistent with the newly formed mineral associations, e.g. the zeolite-bearing parageneses form from elements leached out of basement rocks during smectite and chlorite formation.

Thus, a geochemical effect of the post-impact hydrothermal alteration consists in the concentration of weak basic elements and the depletion of alkali and of high-valency amphoteric elements, which precipitate in cavities as zeolites. From a geochemical point of view, the impact-induced hydrothermal alteration may be compared with sub-alkali metasomatism.

The uniform mineral associations and behavior of chemical elements in the whole volume of the hydrothermal cell provide evidence for rather constant properties of the fluids during impact-induced hydrothermal transformations of the rocks. This conclusion is supported by the analysis of the mineral facies in the distinct zones of alteration. Most informative in this context is the diagram of mineral parageneses in relation to the chemical potentials of Na and H_2O (Fig. 23). This figure presents a solution of a unary multisystem for the six most typical hydrothermal minerals, i.e. smectite, chlorite, Ca–Na-zeolite, laumontite, calcite (anhydrite), and Ca–Fe-garnet. Calcium, Si, and Fe + Mg were chosen as virtual components.

When calculated, 4 invariant points of 6, and 6 univariant lines of 15 turned out to be metastable. The generally observed replacement of mineral parageneses in the vertical direction in impact craters corresponds to a vector passing across

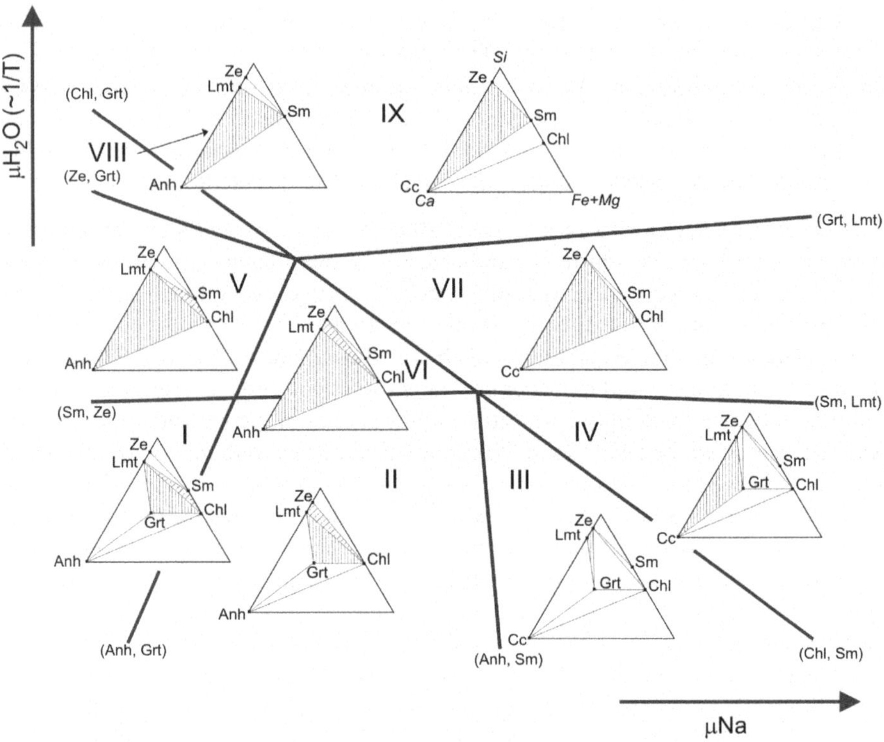

Fig. 23 Parageneses of post-impact hydrothermal mineral associations in relation to the chemical potentials of water (approximately 1/*T*) and sodium (Naumov 2005). The figure presents a solution of a unary multisystem for the six most typical hydrothermal minerals. Calcium, Si, and Fe + Mg are chosen as virtual components. The diagram has been constructed in accordance with topologic regulations, which have been proposed by Schreinemakers and described in detail, by E-an Zen (1966). The diagram has nine stable univariant lines and two stable invariant points. The numbering of fields is a matter of convention. Vertical-hatched areas indicate mineral assemblages that are typical for this field alone, and oblique-hatched areas, typical for two adjacent fields. Compositions of minerals have been accepted by convention as follows: smectite (Sm): $Na_{0.4}(Fe,Mg)_6Si_7{\cdot}9H_2O$, chlorite (Chl): $(Fe,Mg)_9Si_6{\cdot}8H_2O$, Ca–Na-zeolite (Ze): NaCa-$Si_7{\cdot}7H_2O$, laumontite (Lm): $CaSi_4{\cdot}3.5H_2O$, calcite (Cc) or anhydrite (Anh): Ca, and Ca–Fe-garnet (Grt): $Ca_3(Fe,Mg)_2Si_3$

fields I-V-VIII-IX in Fig. 23. Taking into account that the water potential is completely controlled by temperature (H_2O ~ 1/T) at constant pressure, the obtained results indicate only very restricted variations in chemical parameters in the whole hydrothermal cell. Similarly, from *T*-μCO_2 diagrams for calcic minerals (Ryzhenko 1981), the precipitation of either calcite or anhydrite from solutions is governed primarily by the temperature at low PCO_2 values: calcium carbonate is stable at lower than 110–150 °C, whereas calcium sulfate precipitates at higher temperatures. Thus, it is probable that the thermal structure plays the decisive role in the evolution of the mineral associations in space and time.

The generalized order of mineral crystallization is as follows:

in smectite-zeolite zone: saponite/montmorillonite, laumontite/chabazite–calcite, heulandite, analcime–apophyllite, calcite, Fe-montmorillonite, calcite;
in chlorite-anhydrite zone—salite–epidote–chlorite–calcite, anhydrite, laumontite.

For both zones, these sequences confirm the uniform decrease of the temperature of mineral formation during the hydrothermal process.

Inferred from summarized data on alteration mineral distribution, the hydrothermal zonation is not caused by the composition of the target, nor by shock and post-shock thermal metamorphism. The latter is responsible for textural disordering of rocks and controls only the intensity of the hydrothermal alteration. The post-impact thermal field of an impact crater is the main factor controlling mineral formation. The temperature of crystallization of hydrothermal minerals decreases upward as well as outwards from the crater center. The given mineral zones correspond to a medium (chlorite-anhydrite), and a low (smectite-zeolite) temperature facies. The criteria for determining the boundary between both facies is the transition from chlorites to smectites in the temperature range of 160–180 °C, according to data for modern volcanic areas (e.g., Kristmannsdottir 1985; Slovtsov and Moskaleva 1989; Rychagov et al. 1994; Robinson and Santana de Zamora 1999, and others). Temperature assessments for the equilibrium association Ca-heulandite—and laumontite range from 150 to 190 °C (Cho et al. 1987). The given values are considered as maximum temperatures for the hydrothermal transformations in the smectite-zeolite zone. The association of high-silica zeolites with dioctahedral smectites in coptomict gritstones corresponds to temperatures of 80–130 °C (Rychagov et al. 1993), at which montmorillonite is stable in a near neutral environment. Thus, the temperature interval for mineral formation in the smectite-zeolite zone ranges from <100 to 200 °C, and reaches 200–350 °C in the chlorite-anhydrite zone. The latter value is derived from the presence of ferriferous garnets that form in geothermal systems at temperatures between 275 and 350 °C. The absence of corrensite indicates both relatively high temperature gradient in the smectite-zeolite zone and low water-rock ratio. Below, the temperature gradient is apparently very small. The lower Ni/Co ratio in pyrite from the smectite-zeolite zone is evident for lower crystallization temperature and lower sulfur fugacity as compared with chlorite-anhydrite zone.

The chemical parameters of the mineral formation vary within narrow interval of pH and, as a rule, correspond to weakly alkaline and near-neutral (pH = 6–8) environments. It is a sequence of the specific state of host rocks, which consist mainly of shock-derived aluminosilicates and diaplectic glasses, that are readily amendable to leaching. The intense leaching of this material providing the above-mentioned properties of solutions; besides, this is responsible for the oversaturation of solutions with silica, which, in turn, provides favorable conditions for the growth of Fe-smectites and zeolites. Both these minerals dominate in the upper zone of the hydrothermal column. Regular variations in zeolite compositions (increasing Si/Al ratio) and smectites (decreasing Al^{IV} contents) point to some drop in the pH values of the environment upwards. The vertical zoning of zeolite distribution was caused both by the temperature gradient and the drop of partial CO_2 pressure as manifested by the crystallization of Ca–Na-zeolites in association with calcite. The lateral zoning was apparently due to spatial differentiation of acid-basic properties of solutions, which controlled the development of high-silica (heulandite, erionite) either low-silica (Ca-chabazite, analcime) varieties.

The question for the source of the hydrothermal solution can be solved on the basis of a general model for impact-induced hydrothermal circulation (Masaitis and Naumov 1995). This model implies the superficial origin of solutions. Another part of the fluids originates in the shock-induced devolatilization of minerals. The most probable source of Ca, Mg, sulfur, and carbon dioxide are rocks in the peripheral part of the circulation cell where carbonate rocks and evaporites frequently occur in the megabreccias, which fill the ring trough. A preliminary estimate of the isotope composition of sulfur, oxygen, and carbon in some of the hydrothermal minerals (Table 9) does not contradict this suggestion. The relatively light sulfur isotope composition of anhydrite associated with pyrite at deep levels of the section corresponds to the sulfur composition of Permian evaporites in the region, i.e., $\delta^{34}S$ from 6.4 to 14‰ CDT (Vinogradov 1980). These data can be explained by isotope fractionation of sulfate sulfur with a coefficient of 1.007 that corresponds to the temperature of ca. 300 °C.

Stable isotope values in pyrite and calcite from the smectite-zeolite zone confirm unambiguously the superficial source of fluids. All analyzed calcite samples have moderately depleted $\delta^{18}O$ values (less than +26‰, V-SMOW), which are appropriate to normal fresh- or seawater. Carbon isotope values span in a range from strongly (organic matter influenced) and moderately depleted (typical hydrothermal) to slightly enriched (normal marine) carbonates (from −26 to −7‰, V-PDB) (Versh et al. 2006) Mantle or metamorphic derived carbon source of CO_2 can be excluded. Thus, the isotope composition of oxygen and carbon in calcites from veins in the near-roof part of the authigenic breccia strongly suggests a low-temperature origin from a heterogeneous carbon source, which probably includes both carbonates and organic matter.

The principal scheme for the formation of the hydrothermal mineralization is given in the following. Superficial aqueous fluids infiltrate hot rocks of the central uplift and impactites filling the crater, reacting with shock-derived aluminosilicates and impact glass and forming Fe–Mg clay minerals. These fluids become weakly alkaline due to the selective removal of silica, aluminia, and alkalis, plus under neutral environment also of Ca. This is provided by the widespread presence of shock-disordered aluminosilicates and diaplectic glasses. A constant super-saturation of solutions by silica creates favorable conditions for the formation of Fe-smectites and zeolites—the main impact-generated hydrothermal minerals. During ascent in the thermogradiental field, the hydrothermal solutions become more acidic as a result of both the absorption of OH^- ions, as rock-forming silicates are replaced by smectites and chlorites, and because of the increase of the CO_2 solubility due to the decrease of the temperature. This is an important factor against the fall of pressure, which can cause dissolution of hydrocarbonate complexes that lead to boiling of solutions near the surface. Due to the thermal gradient in the authigenic breccia zone, the newly formed mineral associations also show a vertical and horizontal distribution. Silica,

Al, K, and Na are removed, but accumulate partly in the upper zone of hydrothermal *cell by precipitation of zeolites.

Thus, the distribution of hydrothermal mineralization in the Puchezh-Katunki structure is satisfactory explained by the model of a near-surface circulation system, confined to the crater center and controlled by the thermal energy induced by the impact event. The characteristics of the hydrothermal mineralization are typical for impact structures (Naumov 1996, 2002, 2005). The zoning of the hydrothermal mineralization resembling an "abyssal" heat source actually results from the thermal evolution of the impact crater. Many features of the hydrothermal mineral associations at the Puchezh-Katunki, e.g., mineral associations, composition of minerals, temperature zoning, are in keeping with features of hydrothermal associations in volcanic areas. This similarity is the result of the similar thermodynamic and hydrological setting in the near-surface part of the Earth's crust.

4 Thermal Evolution of the Impact Structure

The intense appearance of thermal transformations and hydrothermal alteration in the central uplift of Puchezh-Katunki structure is paid a special attention because it enables to simulate the thermal evolution of a large impact crater based on observed mineralogical effects in addition to speculative physical modelling.

The pre-impact thermal field of the target (crystalline basement and overlying sedimentary cover) was controlled by a geothermal gradient characteristic for stable platform regions. At present, this gradient ranges from 15 to 20° km^{-1} and was most likely the same at the time of the impact in Jurassic.

The impact event changes abruptly the stable thermal regime. A subsequent thermal evolution of the impact structure should be considered with regard to successive stages of cratering.

1. The rocks constituting the present central uplift underwent a shock pressure of no more than 40–45 GPa; more intensely compressed rocks were evaporated, melted and ejected. Thus, residual temperature (immediately after the unloading behind the shock wave front) for rocks, which remained in solid state exceeded an initial (pre-impact) temperature for ca. 900 °C. Taking into account that at the moment of compression these rocks occurred at a depth up to 6–8 km, their post-shock temperatures could be as high as 1000–1100 °C. This temperature was decreasing sharply downward in accordance with shock-wave attenuation; at depths below 10 km, the post-shock temperature gradient became insignificant. Post-shock isotherms under transient crater floor are suggested to be hemispherical following in general the configuration of isobaric surfaces.
2. Additional heating in the subsurface zone of the excavation crater was due to injections of impact melt impregnated with fragments overheated much more than the liquidus temperature (up to 1500–1600 °C or even higher). This heating followed immediately the shock wave and radial displacements of rock masses.

As the distribution of melted material in injected masses was rather irregular, in some places these masses consisted entirely of heated fragments enclosed in thin-crushed material showing (as well as the fragments) signs of viscous flow. This additional heating was irregularly spaced, local, and relatively short-term as far as the volume of injected masses was much less relative to the volume of the heated massif. It is quite probable that the heating was somewhat more intense owing to internal friction in the material during its instantaneous movement between rock blocks.

3. Differently directed fast block displacements behind the shock-wave front resulted in some friction heating along boundaries of blocks, up to local melting and pseudotachylite formation.
4. Elevation of the central uplift at the stage of early modification of the transient crater was accompanied by intense shifting and heating in a conical zone pinching out downward (Ivanov 2008). This heating is probably responsible for the thermal decomposition of minerals and emergence of thermal residual magnetization generating an intense positive magnetic anomaly over the central uplift. In some places, the selective melting occurred; this local melt could mix with still hot impact melt portions.
5. After the central uplift had arisen, a relatively long-term heat anomaly involving a massif of deformed crystalline rocks of conical or ellipsoid shape formed within it. The vertical extension of the massif was up to 5–6 km and the total volume of ca. 1000 km^3 (Masaitis and Naumov 1995; Naumov 1996). The temperature was decreasing from 500–600 °C in the center of the massif to 100 °C at its periphery. The next evolution of the thermal anomaly consisted in temperature equilibration within it and simultaneous general cooling, which was also irregular: it was slower in areas where a thicker (a few hundreds of meters) cover of hot deposited ejecta overlaid the uplift. At the summit of the uplift the ejecta blanket was thin or absent, and therefore, oxidation of hot rocks and some additional heating occurred. During this stage, the conductive heat transfer was predominant (except for the uppermost of the authigenic breccia column). When the temperature in the central part of the massif fell to reach the critical point for water vapor to fluid transition (ca. 400 °C and lower in dependence from pressure), the stage of convective thermal flow started.
6. The last stage of thermal evolution of the Puchezh-Katunki impact structure consisted in the development of hydrothermal circulation. In fact, this circulation started at peripheral parts of the thermal anomaly where temperatures were lower than the critical water point immediately after its stabilization following the completion of the early modification stage. As the massif was cooling, the front of circulation moved toward its centre. Thus, the stages of convective and conductive cooling coexisted for a short time until the convective cell involved all mass of heated rocks.

Intense fluid circulation is provided by both temperature gradient and high permeability within the circulation cell and occurrence of the high-mineralized shallow basins in annular trough and central pit that supply available water reserves due to downward infiltration through relatively loose allogenic breccias and highly fractured authigenic breccia. In the middle of the central uplift, descending flows gave way to ascending ones. Hot water solutions supplied not only into the authigenic breccia, but also into porous breccias and suevites as well as into layers of redeposited material of the crater lake floor.

Three stages of the regressive hydrothermal process might be distinguished (Masaitis and Naumov 1995; Naumov 2002, 2005; Fig. 24). (a) an initial stage, when the isotherms preserve their original configuration (the temperature decreases from the top downward); (b) a main stage after an inversion of the thermal field; the thermal gradients during this stage ranging from about 30 °C km^{-1} in the central part of the thermal anomaly to about 100 °C km^{-1} on its periphery. (c) a final stage, during which thermal gradients are less than 10–30 °C km^{-1} and hot-water circulation takes place in near-surface area only.

The substitution of chlorites by smectites downward in the section is evidence for mineral formation during the initial stage. The creation of the vertical zonation of hydrothermal minerals in the central uplift because of the thermal gradient in the authigenic breccia occurs in the main stage when the circulation system incorporates the maximum rock volume. Finally, zeolite-calcite cement is precipitated within suevites and lithic breccia, and late calcite vein forms.

The time of existence of the hydrothermal system is estimated to be several thousands of years (Naumov 2005). The hot-water circulation hastened the gradual cooling of the impact crater. Finally, the temperature distribution was levelled against the regional geothermal gradient, and the hydrothermal activity decayed.

Thus, it is supposed from the modelling of the thermal evolution of the impact structure that the thermal field in the central part of the sub-crater zone was inverted twice. Firstly, the inversion was caused by the impact-induced origination of a heated rock massif, in which the temperature decreasing downward; secondly, by the preferred cooling of this massif close to surface and reconversion to normal vector direction of the geothermal field.

References

Badyukov DD (1986) Effect of shock waves on main types of rock-forming minerals. Meteoritika 45:122–130 (in Russian)

Basilevsky AT, Ivanov BA, Florensky KP, Yakovlev OI, Feldman VI, Granovsky LB (1983) Impact craters on the Moon and planets. Nauka Press, Moscow 200pp (in Russian)

Bunch TE, Cohen AJ, Dence MR (1968) Shock-induced structural disorder in plagioclase and quartz. In: French B, Short N (eds) Shock metamorphism of natural materials. Mono Books, Baltimore, pp 509–518

Cho V, Maryama S, Liou JC (1987) An experimental investigation of heulandite-laumontite equilibrium at 1000 to 2000 bar P_{fluid}. Contrib Mineral Petrogr 97:43–50

Deer WA, Howie RA, Zussman J (1978) Rock-forming minerals, vol 2A. Single-chain silicates, 2nd edn. Longmans, 668pp

Deer WA, Howie RA, Zussman J (1986) Rock-forming minerals, vol 1B. Disilicates and ring silicates. Geological Society, London, 630pp

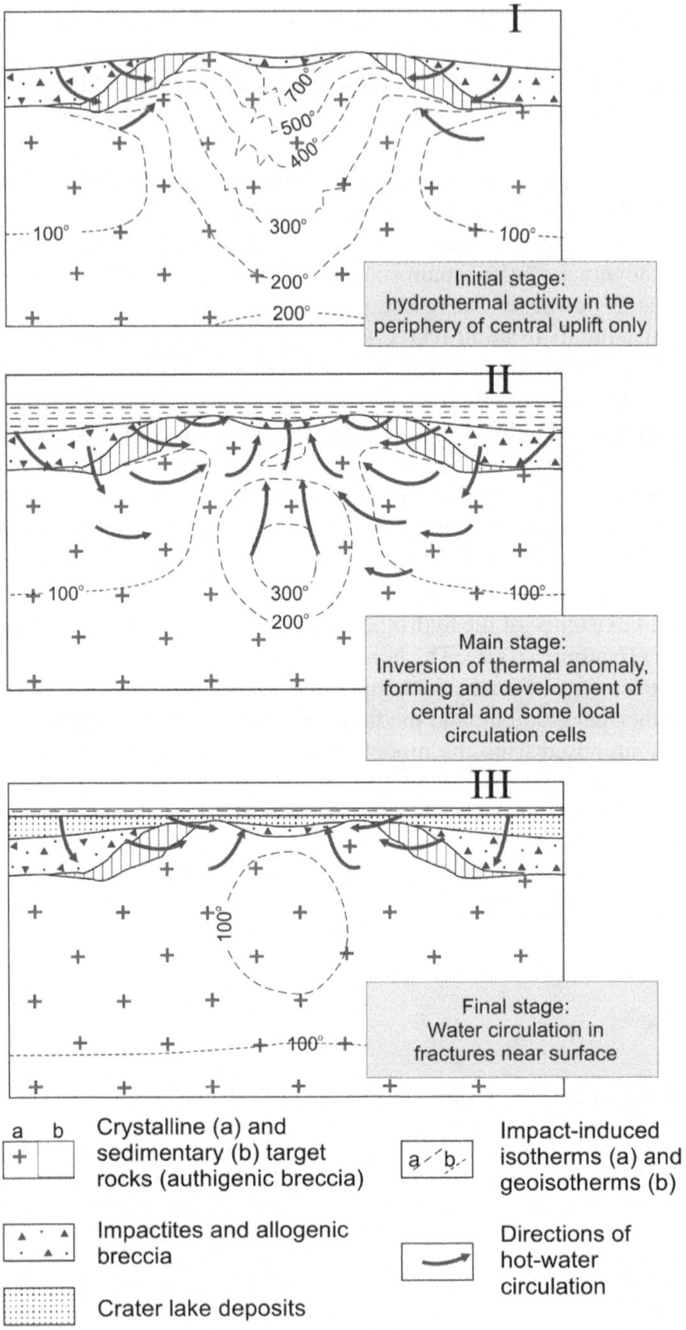

Fig. 24 A general scheme for impact-generated hydrothermal circulation in the Puchezh-Katunki impact crater. Three stages of united regressive hydrothermal process are shown (see text)

Deer WA, Howie RA, Zussman J (2001) Rock-forming minerals, vol 4A. Framework silicates: feldspars, 2nd edn. Geological Society, London, 984pp

Dworak U (1969) Stoswellen metamorphose des Anorthosite von Manicouagan Krater, Quebec, Canada. Contrib Miner Petrol 24:306–347

Feldman VI (1990) Petrology of impactites. Moscow State Univ Press, Moscow 299pp (in Russian)

French BM (1998) Traces of catastrophe: a handbook of shock-metamorphic effects in terrestrial meteorite impact structures. LPI Contribution No. 954, Lunar and Planetary Institute, Houston, 120pp

French BM, Koeberl C (2010) The convincing identification of terrestrial meteorite impact structures: what works, what doesn't and why. Earth-Sci Rev 98:123–170

Gibson RL, Reimold WU, Stevens G (1998) Thermal-metamorphic signature of an impact event in the Vredefort dome, South Africa. Geology 26:787–790

Gnevushev MA, Krasavina TN, Kuznetsova LG (1982) Garnet changes in impact metamorphism. Mineral zb 2:27–31 (in Russian)

Grieve RAF, Masaitis VL (1994) The economic potential of terrestrial impact craters. Int Geol Rev 36:105–151

Harrison WJ, Hörz F (1981) Experimental shock metamorphism of calcic plagioclase (abs.). Lunar Planetary Science XII, Houston, pp 395–397

Hart RJ, Hargraves RB, Andreoli MAG, Tredoux M, Doucouré CM (1995) Magnetic anomaly near the center of the Vredefort structure: implications for impact-related magnetic signatures. Geology 23:277–280

Ivanov BA (2008) Geologic effects of large terrestrial impact crater formation. In: Adushkin VV, Nemchinov IV (eds) Catastrophic events caused by cosmic objects. Springer, Berlin, pp 163–205

James OB, Lindstrom MM, Flohr MK (1989) Ferroan anorthosite from lunar breccia 64435: implication for the origin and history of lunar ferroan anorthosites. In: Proceedings of the lunar planetary science conference XIX, Houston, pp 219–243

Johansson A (1984) Geochemical studies on the Boda Pb-Zn deposits in the Siljan astrobleme, central Sweden. Geologiska Foreningen i Stockholm Forhandlingar 106:15–25

Kristmannsdottir H (1985) The role of clay minerals in geothermal energy research. Clay minerals Nordic symposium. Uppsala, Sweden, pp 125–132

Langenhorst F, Deutsch A (1998) Minerals in terrestrial impact structures and their characteristic features. In: Marfunin AS (ed) Advanced mineralogy, vol 3. Springer, Berlin, pp 95–119

Lindsley DH (1983) Pyroxene thermometry. Am Miner 68:477–486

Lindström MM, Lindström DJ (1986) Lunar granulites and the role of anorthosite norites in the early solar crust. J Geophys Res 91:D263–D276

Masaitis VL (1989) The economic geology of impact craters. Int Geol Rev 31:922–933

Masaitis VL, Danilin AN, Mashchak MS, Raikhlin AI, Selivanovskaya TV, Shadenkov EM (1980) Geology of astroblemes (in Russian). Nedra Press, Leningrad, 231 pp (in Russian)

Masaitis VL, Mashchak MS (1990) Puchezh-Katunki astrobleme: structure of central uplift and transformation of composed rocks. Meteoritics 25(4):383

Masaitis VL, Mashchak MS (1994) Post-impact recrystallization and fusion of target rocks in large impact craters (abs.). In: Vernadsky-Brown microsymposium XXIII, Moscow, pp 84–85

Masaitis VL, Mashchak MS (1995) Terrestrial analogs of lunar granulites (abs.). Lunar Planetary Science XXVI, pp 897–899

Masaitis VL, Mashchak MS (1996) Recrystallization and blastesis of shock-metamorphosed rocks in impact structures. Zapiski Vserossiskogo Mineralogicheskogo Obshchestva 125(4):1–18 (in Russian)

Masaitis VL, Naumov MV (1995) An outline model of hydrothermal circulation in impact craters. Russ. Acad. Sci. Trans. Earth Sci. Sect. 377A(6):22–26

Masaitis VL, Pevzner LA (eds) (1999) Deep drilling in the Puchezh-Katunki impact structure. VSEGEI Press, Saint Petersburg 392pp (in Russian)

McGee JJ (1987) Petrology and precursors of lithic clasts from feldspathic fragmental breccia 67975. J Geophys Res 92(B4):E513–E525

McGee JJ (1989), Granulitic breccia clast and feldspathic melt breccia clasts from North Ray Crater breccia 67975: precursors and petrogenesis. In: Proceedings of lunar planetary science conference XIX, Houston, pp 73–84
Naumov MV (1993) Zeolite mineralization in impact craters. Zap Vses Mineral O-va 122(4):1–12 (in Russian)
Naumov MV (1996) Principal regularities of the post-impact hydrothermal process. Sol Syst Res 30:21–27
Naumov MV (2001) Zeolites in impact craters. In: Galarneau A, Di Renzo F, Fajula F, Vedrin J (eds) Zeolites and mesoporous materials at the dawn of the 21st century. Studies in surface science and catalysis, vol 135, pp 41–48
Naumov MV (2002) Impact-generated hydrothermal systems: data from Popigai, Kara, and Puchezh-Katunki impact structures. In: Plado J, Pesonen LJ (eds) Meteorite impact structures in Precambrian shields impact studies, vol 2. Springer, Berlin-Heidelberg, pp 117–171
Naumov MV (2005) Principal features of impact-generated hydrothermal circulation systems: mineralogical and geochemical evidence. Geofluids 5:165–184
Ostertag R (1983) Shock experiments on feldspar crystals. J Geophys Res Atmos 14(S01):364. https://doi.org/10.1029/JB088iS01p0B364
Robinson D, Santana de Zamora A (1999) The smectite to chlorite transition in the Chipilapa geothermal system, El Salvador. Am Miner 84:607–619
Rychagov SN, Zhatnuev IS, Korobov AD, Kiryukhin VA, Glavatskikh SF (1993) The structure of a hydrothermal system. Nauka Press, Moscow, p 298 (in Russian)
Rychagov SN, Glavatskikh SF, Goncharenko OP (1994) Thermal regime of secondary mineral formation and structure of temperature field in interior of Baransky volcano (Iturup Island). Volcanologija i seismologija 6:96–112 (in Russian)
Ryzhenko BN (1981) Thermodynamics of hydrothermal equilibria. Nauka Press, Moscow, p 192 (in Russian)
Schreier W (1983) Metamorphism and fluid inclusions in the basement of the Vredefort Dome, South Africa: guidelines to the origin of the structure. J Petrol 24:26–47
Seal RR II (2006) Sulfur isotope geochemistry of sulfide minerals. Rev Mineral Geochem 61:633–677
Sinelnikov NN (1956) On cristobalite and its transformation into tridymite. Trans. (Doklady) Acad. Sci. USSR 110:651–654 (in Russian)
Slovtsov IB, Moskaleva GN (1989) Phyllosilicates as probable indicators of PT conditions of hydrothermal process. Volcanologija i seismologija 5:104–110 (in Russian)
Stöffler D (1971) Progressive metamorphism and classification of shocked and brecciated crystalline rocks at impact craters. J Geophys Res 76:5541–5551
Stöffler D (1972) Deformation and transformation of rock-forming minerals by natural and experimental shock process. Contrib Miner Petrol 49:50–113
Stöffler D (1984) Glasses formed by hypervelocity impact. J Non-Crystal Solids 67:465–502
Stöffler D, Langenhorst F (1994) Shock metamorphism of quartz in nature and experiment. I. Basic observation and theory. Meteoritics 29:155–181
Tchukhrov, Bonstedt-Kupletskaya (eds) (1965) Minerals, vol 1. Mir Publishers, Moscow 371pp (in Russian)
Tchukhrov, Bonstedt-Kupletskaya (eds) (1981) Minerals, vol 2. Nauka Press, Moscow 400pp (in Russian)
Tchukhrov, Bonstedt-Kupletskaya (eds) (1983) Minerals, vol 3. Nauka Press, Moscow, p 614 (in Russian)
Versh E, Kirsimäe K, Buchardt B, Naumov MV, Öhman T, Jõeleht A (2006) Mineralogical and stable isotope study of impact-induced hydrothermal carbonate minerals (abs.). In: First international conference on impact cratering in the solar system ESTEC, Noordwijk, The Netherlands, 08–12 May 2006
Vinogradov VI (1980) Influence of sedimentation processes to sulfur isotope geochemistry. Nauka Press, Moscow 192pp (in Russian)

Zen E (1966) Some topological relationships in multisystems of n + 3 phases. 1: General theory; unary and binary systems. Am J Sci 264:401–427

Ziegler VD, Martynenko VV, Rud RF, Fedorov AE, Barkar VG (1988) Formation of a tridymitic texture in siliceous bricks for coke furnaces. Refract Mater 11:21–27 (in Russian)

Impact Event and Subsequent Geologic Evolution

Mikhail V. Naumov, Victor L. Masaitis and Mikhail S. Mashchak

1 Post-Impact Formations

1.1 Crater Lake Deposits

Crater lake deposits consist of two formations, which differ considerably in amount, spatial distribution, composition, and mode of origin; these are: (1) Coptomict gritstones and sandstones (Uzola Formation) and (2) Thin-bedded clays (Kovernino Formation).

The coptomict gritstones and sandstones occur mainly in the central pit and on slopes of the central uplift. In addition, these lithologies are opened by certain boreholes (bhs. 38, 39, 70, 83, Fig. 1) in the southern, eastern and western sectors of the annular trough. At the crater rim, coptomict lithologies occur in a local area at the northwestern sector of the crater (bh. I-64). It is quite possible that they were washed out to a considerable extent during the deposition of the Kovernino Formation.

The coptomict gritstones overlie commonly litho-vitroclastic suevites. The boundary between both lithological types is drawn where signs of rewashing are not visible anymore. Coptomict deposits fill in lows of the surface of the apparent crater floor. So, the hypsometric level of their bottom ranges very considerably. For example, bhs. 44 and 46 in the southwest of the central uplift opened the bottom of the coptomict gritstone at an elevation of +88.1 m, and +98.4 m, respectively, while bh. 26 in the southeast of the uplift, at a depth of −135.5 m (Fig. 1). Major differences in the position of this level are characteristic for the annular trough as well. In the southern part of the latter, bh. 70 penetrated the bottom of the coptomict gritstone at an altitude of −187 m, while bh. 38 in the eastern part, at −343.4 m. The thickness of coptomict deposits varies considerably and irregularly as well. It exceeds 100 m in

M. V. Naumov (✉) · V. L. Masaitis · M. S. Mashchak
A.P. Karpinsky All-Russia Geological Research Institute,
199106 Sredny Pr. 74, St. Petersburg, Russian Federation
e-mail: m_naumov@mail.ru; Mikhail_Naumov@vsegei.ru

V. L Masaitis and M. V Naumov (eds.), *The Puchezh-Katunki Impact Crater*,
Impact Studies, https://doi.org/10.1007/978-3-030-32043-0_5

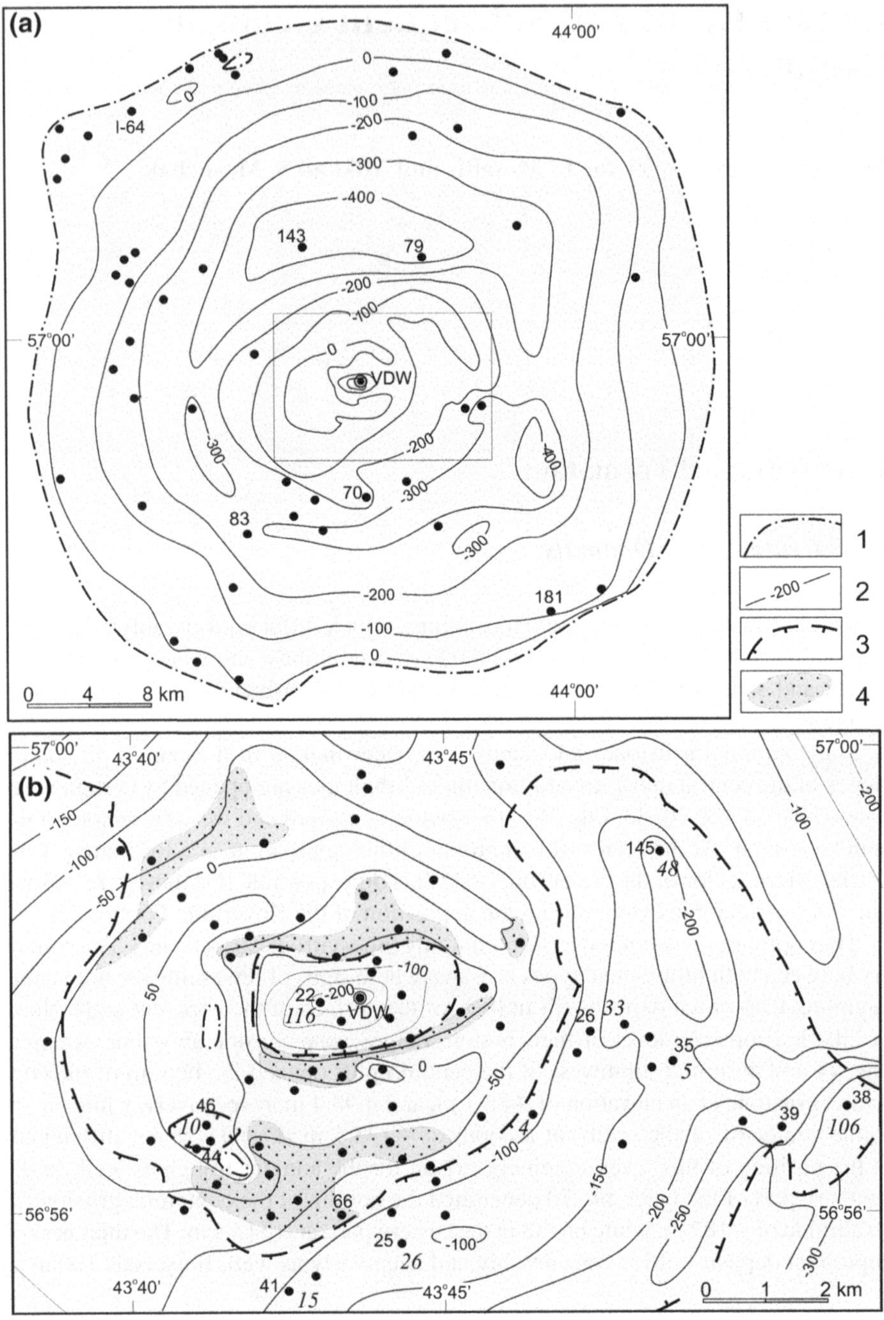
(a)
44°00'
0
-100
-200
-300
-400
I-64
143
79
-200
-100
0
57°00'
57°00'
VDW
-300
-200
-400
-300
70
83
-300
-200
181
-100
0
0 4 8 km
44°00'
1
-200
2
3
4
57°00'
43°40'
43°45'
57°00'
(b)
-150
-100
-50
0
50
-100
-200
22
116
VDW
-100
0
145
48
-200
26
33
35
5
-50
-100
46
10
44
4
66
-150
-200
-250
39
38
106
56°56'
56°56'
25
26
-100
41
15
43°40'
43°45'
-300
0 1 2 km

◀**Fig. 1** Structural scheme of crater lake deposits for the Puchezh-Katunki crater as a whole (**a**) and for the central area in more detail (**b**); the latter is shown by a rectangle at the scheme *A*. 1—limit of crater lake deposits occurrence, 2—contour lines of the base of crater lake deposits and their values (meters above sea level); contour interval is 100 m for (**a**), and 50 m for (**b**); 3—boundary of coptomict deposits areals (hachures point inward to areals), 4—areas of coarse-grained material in the base of Kovernino Formation. The boreholes penetrated the bottom of crater lake deposits are shown by full circles (VDW, by a double circle), boreholes cited in the text are numbered. The thickness of coptomict deposits in certain boreholes is given in italic

the central pit (116 m, bh. 22) and in the eastern part of the annular trough (105.8 m, bh. 38). However, commonly coptomict deposits are a few meters thick and only in individual lows their thickness increase up to a few tens of meters. In whole, the coptomict deposits are distributed over a 340 km^2 (about 25% of the crater area) and reach a volume of 4 km^3, i.e. they surpass the amount of impactites (suevites and tagamites).

The coptomict gritstones and sandstones are dense greenish-grey rocks. They are composed of angular to rounded fragments of various rocks (20–35 vol.%), altered impact glasses (25–50%) and groundmass (30–45%). Lithoclasts are represented by various sedimentary (argillite, shale, siltstone, altered carbonate, clay, sandstone) and crystalline (usually shocked and annealed) rocks in approximately equal proportions. The fragments range from 0.1 to 3 cm across. Relatively large (10 cm and more) fragments of variably altered gneiss, and less frequent amphibolites are found at different levels of the vertical section. Impact glasses are brown or green, porous or massive, commonly montmorillonitized. The matrix is clayey or carbonaceous to clayey, it contains small impact glass particles. Coptomict deposits in the central pit are distinguished by calcite–zeolite cement (Fig. 2) with a pseudo-poikilitic texture. Zeolites are represented by high-silica varieties (Na-heulandite, Na-chabasite, erionite) (Naumov 2002); they contribute up to 10 vol.%, locally peaking as much as 30 vol.%. Formation of zeolites is provided of the enormous amount of products of leaching of shock-disordered aluminosilicate matter.

The chemical composition of the coptomict gritstones is close to that of suevites (see Tables 3.5, 3.10, 3.11). Some differences are lower contents of magnesium and iron, higher Fe_2O_3/FeO ratio (6.0 compared to 2.2 for the suevites), and higher strontium and volatile contents.

Upwards in the section, coptomict deposits show an uniform change of their structure. The fragments size are decreasing, the roundness of the fragments as well as the contribution of both impact glasses and groundmass increasing. Thus, close to the top, some lenses (up to 4 m thick) of sandstone, siltstone, and clay appear within the Usola Formation to mark a gradual transition to the overlying Kovernino Formation.

In addition, coptomict gritstones and sandstones with a cement of hydrothermal origin build clastic dykes up to tens of centimeters thick. Such dykes were intersected by the VDW in suevite, polymict allogenic breccia, and in underlying crystalline rocks (the lowest thin dyke is opened at a depth of 573 m in VDW section), i.e. in

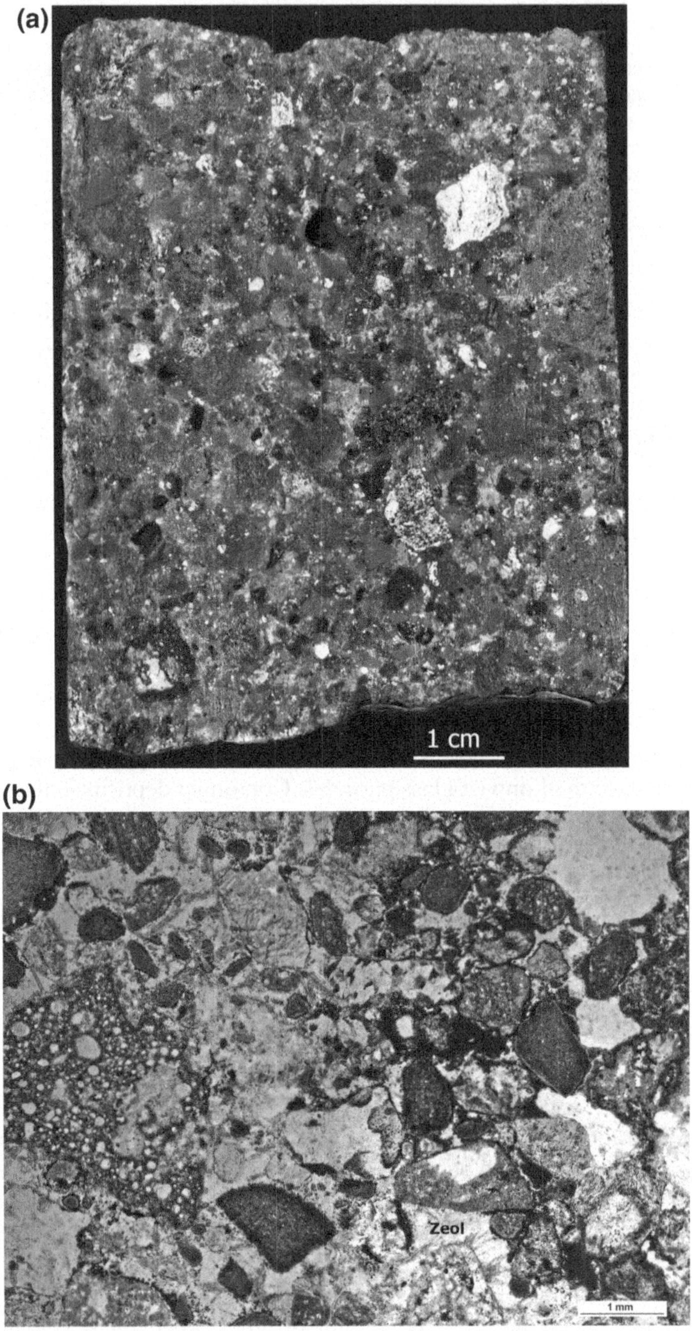
(a)
1 cm
(b)
Zeol
1 mm

◀**Fig. 2** Images of coptomict gritstone with the cement of hydrothermal origin from the central pit (VDW section). **a** A core sample, depth 365 m. **b** Photomicrograph of a thin section, depth 343 m, plane-polarized light. Angular-rounded fragments of claystones and siltstones (dark-colored), of crystalline rocks (light-colored) including shocked ones (light-gray), porous impact glasses (at the left of the picture) etc. are cemented by calcite and zeolite

a depth of more than 200 m below the lower boundary of the coptomict rocks. This observation evidences that coptomict gritstones formed immediately after suevite and allochthonous breccia deposition when deep open fractures existed in the apparent crater bottom. There, the not-lithified material of products of suevite and microbreccia destruction penetrated together with mineralized water.

The bulk of the crater lake deposits regards to the Kovernino Formation. It consists mostly of horizontally bedded clays. The area of their distribution is of about 1550 km^2 and an estimated volume is of 290 km^3. The present outer limit of its occurrence coincide with the outer edge of the annular trough (Fig. 1). The Kovernino Formation overlies coptomict gritstones, allogenic breccias, suevites, and (at the top of the central uplift), shock-transformed crystalline target rocks. Within the crater, this formation is absent only in the highest points of the central uplift (of no more than 0.5 km^2 total area) (Fig. 1), where Miocene sands lie directly over brecciated Neoproterozoic argillite.

The thickness of the Kovernino Formation is determined mainly by the topography of the apparent crater bottom. Thus, its maximum drilled thickness (340 m in bh. 143; 331 m in bh. 79; both boreholes did not reach the base of Kovernino Formation) occurs in the annular trough. Bogorodskaya and Tumanov (1980) assumed the full thickness peaks 450 m. In the central pit, the thickness of the Kovernino Formation is also high (203 m in the VDW section); so, crater lake deposits (including coptomict ones) in central pit and annular trough are close in their thickness.

The Kovernino Formation has a very uniform section all over the crater. It consists of dark-grey, brown-grey, and grey clays with interlayers of siltstones, locally with limestones. Siderite concretions occur occasionally. Clays are composed mainly of kaolinite and montomorillonite. They are frequently silty or calcareous, thin-laminated, in places foliated, with admixtures of silty or sandy material, mica flakes, pyrite or plant detritus along bedding surfaces. Differently colored clay beds usually range from 0.1 to 10.0 m thick. The siltstones are commonly grey, light-grey and greenish-grey, they showing frequently transitions into fine-grained quartz sandstones. The thickness of siltstone interlayers and lenses varies from 0.5 mm to 20 cm but in rare cases, it reaches 6.7 m.

Two members are distinguished in the section of the Kovernino Formation. The upper one, which is 2 to 3 times thicker than the lower, contains lenses and interlayers of limestone (up to 2 vol.%) and (in the southern and southeastern parts of the crater) brown coals (up to 0.4 m thick); siltstones and sandstones occur in places as well (no more than 2 to 3%). In the lower member, limestones are commonly lacking, but siltstone and sandstone make up to 20–30% of the total volume. At peripheral parts of the annular trough, the upper member is thinning out sharply, and on the slope of the trough as well as on the crest of the central uplift, it often is lacking. From palynological data, the lower member had been dated by the late Bajocian, and the upper one, by the early Bathonian (Bogorodskaya and Tumanov 1980). In general, the Kovernino Formation is very poor in fossils except for rare debris of the foraminifer *Ammodiscus* sp. (Fadeeva 2000; Kirkov et al. 2019).

Locally in the central pit, on the slopes of the central uplift, and (very seldom) on the outer slope of the annular trough, layers and lenses from a few centimeters up to 18 m thick (bh. 66, southern slope of the uplift) filled with coarse-grained material derived from impact and target rocks occur at the base of the Kovernino formation (Fig. 1). The composition and granulometric size of this material vary spatially within the crater. In the area of the central uplift, well-rounded sandy material derived from gneisses occurs together with detrital gneisses fragments; in addition, gritstones or conglomerates consisting of Neoproterozoic claystones or siltstones occur. This indicates the existence during the early Kovernino time of inliers of Precambrian or Neoproterozoic rocks raise over the crater lake bottom. Variously rounded fragments and pebbles of marls, limestones and sandstones are derived from breccias in the ring terrace, have been recorded in the lowermost Kovernino clays at the edges of annular trough. In addition, some signs of intra-formational washing out of Kovernino deposits have been found in places. These signs are the presence of rounded and semi-rounded fragments of grey and dark-grey dense clays and siltstones within the same clays.

1.2 Overlying Deposits

The Puchezh-Katunki crater is entirely overlain by later deposits, which include a sequence of shallow marine sediments ranging from Middle Jurassic (upper Bathonian to Callovian stage) to Lower Cretaceous in age, and Miocene and Quaternary continental deposits. Only the variegated breccia in the western sector of the outer ring terrace are projecting from under overlying formations due to removing of the latter by Volga River.

On the present surface, the marine Jurassic and Cretaceous deposits are confined in outline to the contours of the crater depression, but in the north and, locally, in the northeast spread beyond it. They are also lacking in the area of the crest of the central uplift and the central pit (Fig. 3).

The distribution of Middle Jurassic-Upper Cretaceous marine sediments is controlled to a considerable degree by the crater depression. It is clearly demonstrated by relief of its base (Fig. 3). The deepest hypsometric marks of the base of the Callovian (−45 m, bh. 79) as well as the maximum thickness of overlying deposits (162 m, bh. 143; see Fig. 3.12) are recorded in the northern part of the annular trough similarly to the Kovernino Formation. Toward the edges of the trough, this base is rising up to +95.0 m at the crater rim (bh. 129) and to +107.2 m in the area of the central uplift (bh. 46). In the central uplift area (including central pit), the thickness is commonly 40–70 m, This indicates that the apparent crater was not fully filled in by

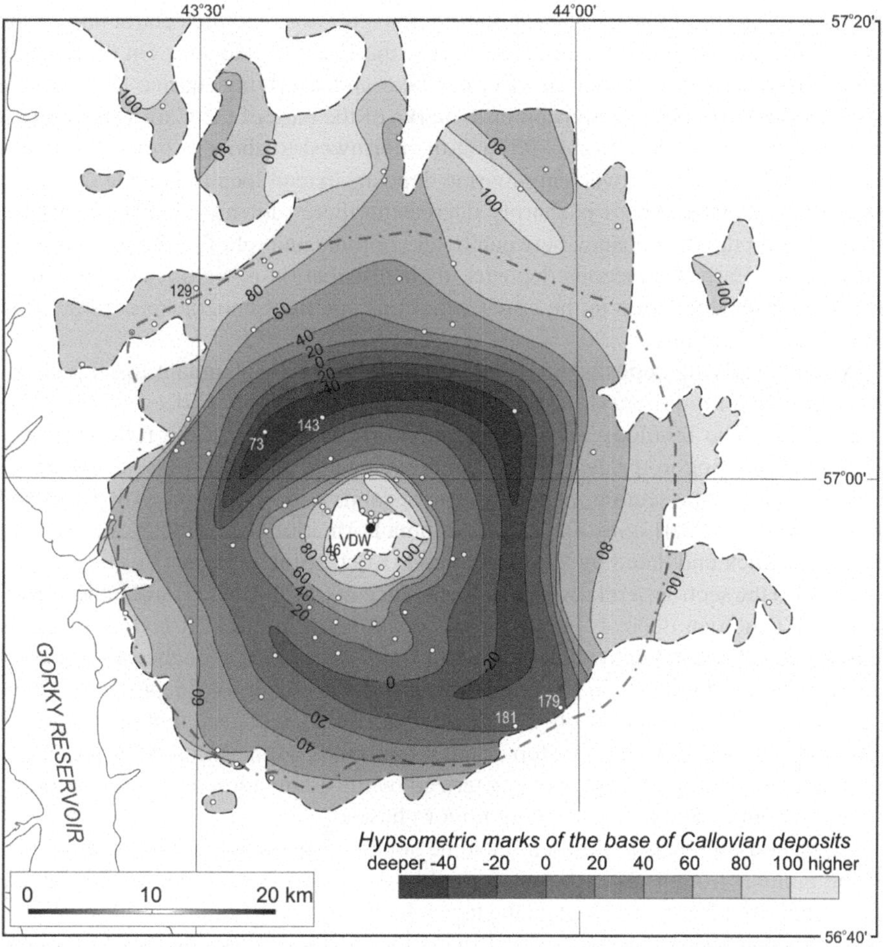

Fig. 3 Hypsometric map of the base of Callovian deposits in the area of the Puchezh-Katunki impact structure. The map compiled based on drilling data as well as results of mid-scale mapping (Blom et al. 1957; Tumanov and Bogorodskaya 1975; Kochergina et al. 2018; Kirkov et al. 2019); data of S.N. Bobrov (unpublished Tech. Report of Volgageologija 1985) are used as well. Contour line interval, 20 m. A limit of spreading of Middle Jurassic–Lower Cretaceous marine deposits is shown by a dashed line, limit of crater lake deposits occurrence, by a dashed-dot line. Boreholes opened the base of the Callovian are shown by open circles, the VDW, by black circle. Boreholes cited in the text are numbered

sediments until the Middle Jurassic marine transgression, and the central part of the annular trough continued to sink even during the Early Cretaceous. On other hand, some differences in distribution of crater lake and overlying marine deposits are observed. In particular, the hypsometric marks of the base of the Callovian at edges of crater depression vary from +95 m at the northwest to about −3 to at the southeast (bh 181). Although overlying marine deposits extend locally outside the crater depression, their thickness is sharply decreasing there (down to 3–5 m in places). Inferred from the discrepancy between outer contours and relief of the base of Middle Jurassic–Early Cretaceous deposits, their distribution is significantly modified by the intense denudation in the Late Cretaceous and the Cenozoic as contracted to crater lake formations.

All the overlying deposits are dated with confidence by abundant fossil associations, mostly foraminifera (Bogorodskaya and Tumanov 1980; Fadeeva 2000; Mitta et al. 2012). The lithology of deposits varies in the section in the following way (Tumanov and Bogorodskaya 1975; Kochergina et al. 2018; Kirkov et al. 2019).

In the base of the section, light-grey quartz sands with subordinate siltstones and thin interlayers of dark-grey clays (Khokhloma Formation) occur. They are no more than 20 m thick and dated by Late Bathonian (Kochergina et al. 2018).

Upward the section, a relatively thick sequence of Callovian, Oxfordian, Kimeridgian and Tithonian deposits stages occurs peaking 96 m thick. It is dominated by thin-bedded clays; at the lowermost (Lower Callovian) of the sequence alone, sands continue be predominant in the southern part of the Kovernino depression. Clays are light-, dark-, or greenish-grey, in places calcareous, with rare thin interlayers of siltstones and marl; close to the top of the sequence, phosphate nodules appear. The mineral composition of clays is characterized by the domination of montmorillonite, while hydromica and kaolinite being minor phases.

Cretaceous deposits are present in some smaller separated areas in the axial part of the annular trough. They lie conformably over the Upper Jurassic and include sediments from Berriasian to Aptian stages. Its maximum thickness is 74 m. Within the Cretaceous sequence, some local ravinements occur fixing the position of the region at a peripheral zone of a marine basin. In the lower part of the section (Berriasian to Valanginian, up to 17 m thick), fine-grained, grey to greenish-grey, quartz-glauconite sandstones with phosphate-clayey cement are predominant; rare interlayers and lenses (up to 1.7 m thick) of siltstone and clays with phosphate nodules occur as well The bulk of the sequence (from Hauterivian to Aptian) consists of dark-gray and black clays with rare lenses of silt, sand, and marl.

Continental Miocene deposits (Tortonian stage) are mainly developed in eastern, central, and southwestern sectors of the structure. They overlie both Jurassic-Cretaceous formations and impact breccias. The Miocene is represented by alluvial and lacustrine facies of total thickness up to 43 m. It consists of light-gray quartz sands with thin (0.2–0.3 m) clay interbeds.

The Quaternary deposits (Lower Quaternary to Recent) are made of sands, loams, sandy loams and clays of various genetic types (Kirkov et al. 2019). They form a continuous cover from a few to 55 m thick. The main body of these sediments consists of Middle Quaternary fluvioglacial sands and morainic loams, sandy loams with pebbles and boulders of the Dnieper Glaciation. Middle Quaternary alluvial and fluvioglacial sands, sandy loams, loams of the Moscow Glaciation and Upper Quaternary and modern alluvial sands, loams, clays of flood-plain and terrace deposits are widespread in valleys of major river channels. Alluvial sands, loams, and gravel fill the Paleo-Volga valley buried under Middle Quaternary and recent sediments. Lower and Middle Quaternary sands, loams, clays occur throughout the area as restricted fields. Locally, marsh peat bogs spread out with small interlayers of clays.

2 Dating of the Impact Event

Despite of the detailed exploration (palynostratigraphic studies, radioisotopic dating technique, paleogeographic analysis etc.), the time of the Puchezh-Katunki impact event is not substantiated with confidence. The older age limit for the impact structure is determined by the presence of fragments of Induan (Lower Triassic) claystones in allogenic breccias, while the younger one, by the age of crater lake deposits (Kovernino Formation), which are conformably overlain by Middle Jurassic (Upper Bathonian and Callovian) marine deposits. Thus, from geological evidences, the Puchezh-Katunki arose between 247 and 167 Ma.

The Triassic age of the impact structure should be rejected due to its good preservation taking into account that Middle to Late Triassic was an epoch of the intense denudation in the central part of the Russian Plate (Nalivkin and Yakobson 1985). Whereas Induan (Lower Triassic) and partially Lopingian (Upper Permian) sediments south of the impact structure were removed completely, the thin (no more than 100 m thick) cover of variegated allogenic breccias in the ring terrace preserved its thickness along the full crater perimeter.

Taking into account the gradual transition from impact formations to crater lake deposits, a special attention is paid to the dating of the latter. These deposits are lacking in fauna fossils; thus, palynological data take on special significance.

A detailed palynostratigraphic analysis of different formations in the area under consideration was performed by G.E. Donskova during the geological mapping of 1:200,000 scale (see Bogorodskaya and Tumanov 1980). This study showed that numerous samples obtained from both allogenic breccias and Neoproterozoic brecciated claystones and Devonian carbonates contain spore and pollen assemblages. The study of core from a continuous interval (174.0–329.3 m) from bh. 17 in the northeastern part of the central pit showed that similar spore and pollen assemblages develop throughout the section including the lowermost of Kovernino Formation, suevite, allogenic breccia and brecciated Neoproterozoic claystones. The complete miospore assemblage is as follows: *Lycopodium marginatum* K.-M., *L. subrotundum* sp., *L. cibotium* sp., *Osmundia* sp., *O. longirimosa* Klim., Matoniceae, *Salvinia pelpurchra* Bolch., *Picea* sp., *P. exiliodes* Bolch., *P. singularia* Bolch., *P. complanataeformis* Bolch., *P. gigantissima* Bolch., *P. mesophytica* Pokr., *Piceites expositus* Bolch., *Podocarpus major* (Naum.) Bolch., *P. triacoca* (Naum.) Bolch., *Protopodocarpus* sp., *Pseudopicea variabiliformis* (Mal.) Bolch., *P. magnifica* Bolch.,

Paleoconiferus asaccatus Bolch., *Ginkgo* sp., *G. typica* (Mal.) Bolch. Although ratios of different miospores vary highly, pollen of gymnosperus being predominant generally. Among the gymnosperms, conifers prevail while pteridophytes are predominant among spores. Such an assemblage is interpreted by G. E. Donskova to be characteristic of Bajocian deposits in the East European Platform.

An abundant pollen and spore assemblage has been established in clays of Kovernino Formation. Spores commonly make up from 5 to 21%, but in places, up to 40%. The most abundant are spores of pteridophytes: *Coniopteris* sp,. *C.trigueta* (Naum.) K.-M., *Leiotriletes typicus* Naum., *L. glaber* (Naum.) Waltz., *Cibotium junctum* K.-M., *Gleichenia* sp., *G. delicata* Bolch., Spores of lycopods and equicetaceous plants are also very diverse: *Lycopodium subrotundum* K.-M., *Selaginella* cf. *cepulinaeformis* K.-M., *S. granata* Bolch., *Osmunda tuberculata* K.-M., *Adiantum mollis* Bolch., *A. glaber* Bolch., *Brochotriletes vulgaris* (Naum.,), *Matoniaceae, Cyathea* sp. The most of assemblages are dominated by pollen of the conifers, which contribute from 63 to 90%. Predominant are *Picea* (12–47%): *Picea exiloides* Bolch., *P.* aff. *singulatae* Bolch., *P. complanatiformis* Bolch., *Pseudopicea variabiliformis* (Mal.) Bolch., *Piceites latens* Bolch., and *Pinus* sp. (10–34%): *Pinus pernobilites* Bolch., *P. subconcinua* (Naum.) Bolch., *P. vulgaris* (Naum.) Bolch., *Pseudopinus textilis* Bolch. Also occur *Podocarpus cretacea* (Naum.) Bolch., *Podozamites* sp., *Bennettites dilucidus* Bolch., *Orbicularia assymetrica* Mal. In addition, pollen of *Gingkoales* (*Ginkgo* sp. and *G. parva* (Naum.) Bolch.) contribute from 5 to 13%. In certain intervals from the lower and middle parts of Kovernino Formation, pollen of genera *Sciadopitus, Brachyphyllum*, and *Pagiaphyllum* are abundant to contribute up to 63%. G. E. Donskova has ascribed this assemblage to the Bathonian Stage.

The spore and pollen assemblages from the lower member of Kovernino Formation are also dominated by pollen of gymnosperms, but the composition of miospore association is distinct there. It includes *Coniopteris* sp., *C.* cf. *divaricuta* Naum., *Cibotium junctum* K.-M., *Leiotriletes* sp., *Lycopodium subrotundum* K.-M., *L. typicus* Naum., *Chomotriletes anagrammensis* K.-M., *Osmundia* sp., *Adiandtum* sp., *Picea* sp., *P. mesophytica* Pokr., *P. exiloides* Bolch., *P. singulata* Bolch., *Pinus* sp., *Protopinus* sp., *P. vastus* Bolch., *P. latebrosa* Bolch., *Podocarpus* sp., *P. triacoca* Bolch., *Paleoconiferus* sp., *P. asaecatus* Bolch., *Bennettites dilucidus* Bolch., *Orbicularia* sp., *Ginkgo typica* (Mal.) Bolch. According to G. E. Donskova, the listed series of miospores indicates an older age; so, it can be assigned to the Bajocian.

Thus, the above palynological study showed that different formations including brecciated target rocks, allochtonous impact breccia, and the lowermost of crater lake deposits, contain the same spore-pollen assemblage, which is characteristic for the Bajocian stage of the central part of East European Platform. The presence of this assemblage in blocks of disturbed Neoproterozoic and Devonian rocks has to be explained by the infiltration of water from the crater lake. At the same time, no Aalenian nor Early Jurassic spores and pollen have been found in the samples. The results indicated unambiguously to the Bajocian age of the Puchezh-Katunki impact structure (ca. 167–171 Ma; Cohen et al. 2013).

Table 1 K–Ar dating of tagamites and impact glasses from the Puchezh-Katunki impact structure

Sample numbers	Rocks	K, %	Ar, %	rad. ^{40}Ar 10^{-6} cm^3/g	Age, Ma
14/253.3[a]	Impact glass	2.32	31.3	0.011	183 ± 5
VDW/753.0[b]	Tagamite	1.86 ± 0.02	8.7	15.3	200 ± 3
VDW/753.0/1[b]	Tagamite	1.86 ± 0.02	11.1	14.8	195 ± 3
44/269.9[b]	Tagamite	0.40 ± 0.01	33.3	3.01	184 ± 3
44/269.9[b]	Tagamite	0.40 ± 0.01	31.3	3.05	188 ± 3

[a] Age identified by L.V. Kopylskaya (Voronezh State University (Bogorodskaya and Tumanov 1980))
[b] Age identified by E.M. Kolesnikov (Moscow State University)
Sample names show the borehole number and depth (in m)

However, radioisotopic dating of tagamites and impact glasses from the Puchezh-Katunki structure (Masaitis and Pevzner 1999) was in contradiction with the palynological age. K–Ar ages of tagamites from both allochtonous breccia and uplifted crystalline basement range from 183 ± 5 Ma to 200 ± 3 Ma (Table 1) that covers the time interval from Hettangian to Toarcian (Lower Jurassic); the average value is 190 Ma. The discrepancy was explained by contamination of the investigated impact melt samples by fragments of Early Precambrian target lithologies causing older apparent ages. Thus, the palynological age for the impact event has been favored (Masaitis and Pevzner 1999).

However, palynological dating of the Kovernino Formation was later disputed. E. M. Rumyantseva (unpublished data from the Technical Report on large-scale mapping of sheets O–38–99, O–38–111) assumed an Early Jurassic age of spore-pollen assemblages from the lower Kovernino Formation.

Isotope U–Pb dating of zircons (including grains with PDFs) from shocked and thermally-metamorphosed target rocks (Naumov et al. 2015) showed that high-level shock metamorphism ($P > 35$ GPa) causes some loss of radiogenic Pb from metamict zircons, but no full erasing of previous isotopic information is occurring. Thus, a separation of specific phases newly formed within zircone grains due to shock and post-shock recrystallization should be performed to date the impact event by zircone measurements.

Recently, a new isotope dating of tagamites has been performed using the ^{40}Ar/^{39}Ar techniques (Holm-Alwmark et al. 2019). The measurement of five samples from tagamite injections within authigenic breccia of the central uplift (VDW section) resulted in one plateau date and four mini-plateau dates that range from 190.3 ± 0.6 Ma to 194.5 ± 0.7 Ma (Table 2). At the same time, the spectrum from a partly glassy and microcrystalline sample (VDW-931) reveal a slight hump-shape with younger step ages (~166–178 Ma) in the low- and high-temperature extraction range and older apparent ages of ~190–195 Ma in the mid-temperature steps. This can be explained by the irradiation process or resulted from alteration of tagamite (Holm-Alwmark et al. 2019).

Table 2 Summary results of $^{40}Ar/^{39}Ar$ dating of tagamites from the Puchezh-Katunki impact structure (from Holm-Alwmark et al. 2019)

Sample	Apparent age (Ma ± 2σ)	Spectrum characteristics	Total ^{39}Ar released over (mini-) plateau (%)	MSWD	Inverse isochron age (Ma ± 2σ)	MSWD	$^{36}Ar/^{40}Ar$ intercept
VDW/931		No plateau					
VDW/1018	196.3 ± 0.8	Mini-plateau	58.1	0.18	193.2 ± 2.0	0.41	298 ± 6
VDW/1144		No plateau					
VDW/1218	193.7 ± 1.1	Plateau	79.3	0.41			
VDW/1251	192.1 ± 0.6	Mini-plateau	58.0	0.38	190.1 ± 1.3	0.83	294 ± 7

Sample names show the borehole number and depth (in m)

Thus, the $^{40}Ar/^{39}Ar$ dating indicates the Sinemurian age for the Puchezh-Katuki impact event to confirm previous K–Ar ages. To adjust radioisotopic age in accordance with earlier palynological data, the latter have been re-evaluated by S. Lindström (Holm-Alwmark et al. 2019). According to her conclusion, the listed by G. E. Donskova spore-pollen assemblages should be attributed to a maximum–minimum age of early Pliensbachian–early Aalenian (ca. 191–170 Ma), but are most likely restricted to the Pliensbachian (ca. 191–183 Ma). This is based primarily on the obtained composite range for *Paleoconiferus asaccatus*, which has a common occurrence in Hettangian to early Pliensbachian strata, with only rare occurrences in late Pliensbachian to early Aalenian strata, and is absent from younger Jurassic assemblages. The restriction to the Pliensbachian is based on the co-occurrence of *P. asaccatus* and *Lycopodiumsporites subrotundus* (Holm-Alwmark et al. 2019). Assemblages from the middle and upper Kovernino Formation described by G.E. Donskova are assumed to be assigned to early Toarcian due to presence of *Gleicheniidites* spp. and *Leptolepidites* sp.

The new palynological analysis of 6 samples from the lower Kovernino Formation and coptomict gritstone performed by S. Lindström, resulted in the recognition of three spore-pollen assemblages. The older one is obtained from a coptomict gritstone. It is dominated to 25% by bisaccate pollen and attributed to an early Sinemurian age due to the presence of *Cerebropollenites macroverrucosus* (Holm-Alwmark et al. 2019). The lowermost Kovernino Formation contains a similar spore-pollen assemblage; however, the presence of *Manumia delcourtii* and *Ischyosporites variegatus* indicates an age younger than late Pliensbachian. The palynological assemblages from the lower Kovernino Formation indicate a depositional age younger than the late Pliensbachian, based on the first occurrences of *Manumia delcourtii, Ischyosporites variegatus*, and the marine dinoflagellate cyst *Mendicodinium groenlandicum*. The absence of late early to late Toarcian markers, e.g., *Callialasporites* spp., *Sestrosporites pseudoalveolatus* and *Staplinisporites caminus* may indicate an early Toarcian age. All three assemblages contain various amounts of reworked Triassic and even Permian spores and pollen (Holm-Alwmark et al. 2019).

Moreover, in both coptomict gritstone and lower Kovernino formation marine acritarchs including *Leiosphaeridia* sp. *Micrhystridium* spp., and *Cymatiosphaera* sp., as well as dinoflagellate cysts assigned to *Mendicodinium groenlandicum* and a finely granulate form of *Mendicodinium,* are found.

Thus, the new palynological analysis of a limited number of samples combined with the re-evaluation of the previously reported palynological assemblages, indicate a Pliensbachian to early Toarcian age for crater lake sediments. In addition, the presence of marine acritarchs and rare dinoflagellate cysts indicate a marginal marine, probably lagoonal setting. This is surprising because of Early Jurassic deposits are lacking in the central part of East European Platform.

Nevertheless, the Early Jurassic age of Puchezh-Katunki impact structure could not be accepted finally because of it comes into contradiction with paleogeographic and geological evidences. The discrepancy between isotopic (Sinemurian) and palynostratigraphic (possibly Pliensbachian to Toarcian) ages remains pendant: from geological observations, crater lake sediments start to deposit immediately after the impact. The palynological data are too ambiguous; moreover, the revealing of marine fossils in crater lake deposits aggravates the uncertainty because marine environments in Moscow syneclise (and in the whole of East European Platform except some marginal areas) were absent for sure during Early Jurassic and Aalenian age; a marine transgression appeared at the end of the Bajocian (Sazonova and Sazonov 1967; Petrov and Kirikov 2006; Zorina 2012). In addition, only local discontinuity occurs between the latter and overlying marine sediments of Upper Bathonian to Aptian that are dated by various fossils (Fadeeva 2000; Mitta et al. 2012; Kirikov 2002). If the Kovernino Formation of about 400 m thick is assumed to start in Sinemurian (190–192 Ma), it should accumulate over 24 Ma (an averaged depositional velocity is 0.02 mm/year). This is incredible for local, mainly lacustrine sediments. Moreover, in the Bathonian the crater was not filled with sediments (hypsometric level of Callovian deposits in Kovernino depression is lower than outside of the crater rim—Fig. 3). In the last version of the unified regional stratigraphic chart for the Jurassic of East European Platform (Mitta et al. 2012), a Bathonian age of Kovernino Formation is not doubted.

Thus, a precise age of the Puchezh-Katunki crater remains obscure. This is caused by some equivocation of both isotopic and biostratigraphic results. From isotope (K–Ar, Ar–Ar) dating, the age is Early Jurassic (Sinemurian), from geological and paleogreographic data, it is rather Middle Jurassic (Bajocian); palynostratigrafic data are too uncertain and contradictory; however, the modern taxonomy indicates rather Early Jurassic age for crater lake deposits.

The data on age estimations for the Puchezh-Katunki impact structure are summarized at the Fig. 4. To date, an Early Jurassic (Sinemurian or Early Pliensbachian; ca. 190–192 Ma) age for the Puchezh-Katunki impact structure is assumed to be

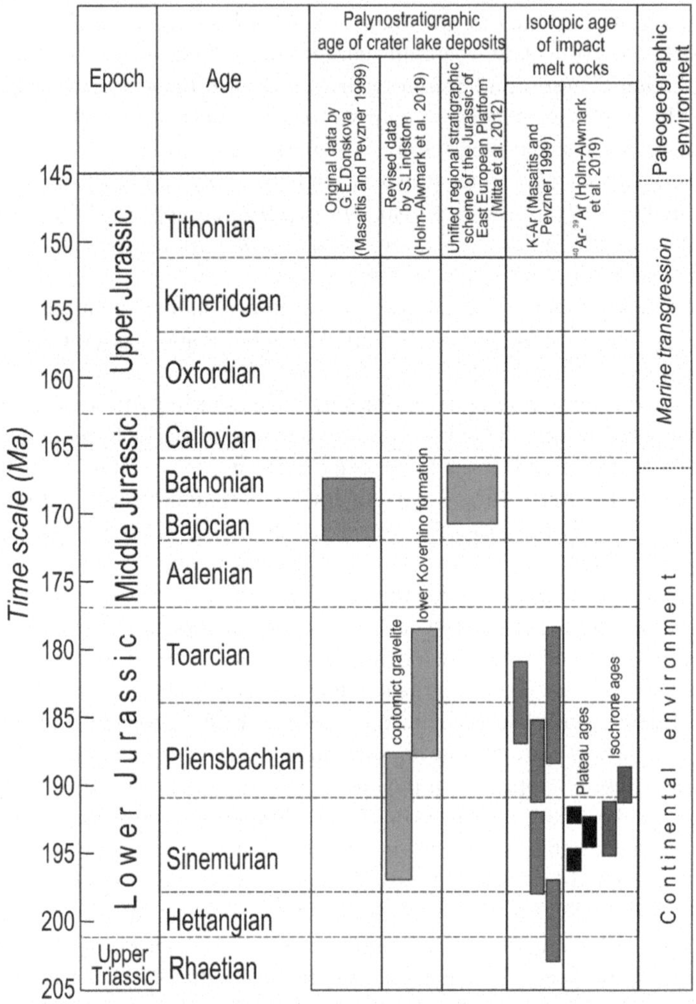

Fig. 4 Summary of dating results for the Puchezh-Katunki impact structure (modified from Holm-Alwmark et al. 2019) Are shown results of earlier K–Ar and recent $^{40}Ar/^{39}Ar$ dating (the boxes display age ranges including errors) and results of palynostratigraphic analysis, which show the most likely age ranges for taxa. Paleogeographic environment during Jurassic for the East European Platform is given in addition

more warranted. Nevertheless, more reliable evidences for the Early Jurassic age of the crater lake deposits including Kovernino Formation should be obtained to solve at last this question.

3 Late Modification of the Impact Structure

Inferred from paleogeographic reconstructions, the impact site occurred at an accumulative plain with numerous lagoon and lake basins. Just after impact, the arisen impact crater was supplied with saline lagoon and fresh lacustrine waters. The major portion of these waters were absorbed by the huge body of high-permeable impact breccias to provide a fluid source for the long-term hydrothermal circulation, which affected a basal portion of crater lake deposits as well (Masaitis and Naumov 1995; Naumov 2002). Therefore, local basins occurred only in lows of the crater. The basins occupied both annular trough and local depressions on the slopes of the central uplift as well as in the central pit, which was fully enclosed by a rocky rim made up by crystalline bedrock. In the basins, coptomict deposits (mainly gritstones) accumulated due to intense destruction of the weakly consolidated impact ejecta blanket (mainly suevites and microbreccia). Calcite–zeolite cement of the coptomict gritstones in the isolated central pit indicates that this basin was saline; on the contrary, in the annular trough coptomict gritstones cemented mainly by clayey and sandy-clayey material.

After the subsidence of the crater bottom due to consolidation of breccia, the water level in the crater was somewhat elevated. Different movements of blocks of the central uplift and its partial destruction resulted in both smoothing of the relief and summary subsidence of the uplift followed by the inundation of the central pit with water from the annular crater lake. So, an indivisible basin filled up the whole of the crater including the circular ridge of the central uplift, and the Kovernino Formation began to deposit. The sourceland included a variegated breccia blanket at the ring terrace, remnants of suevite and microbreccia at outer slopes of the annular trough, and isolated cliffs of Precambrian and Neoproterozoic rocks in the central uplift. At this time, the topography of the provenance was planed to a considerable extent; the surface was covered with soil and vegetation including vast forestlands. Therefore, the crater lake was filled with fine, clayey material. Only around the central uplift where bedrock snouts remained in places, some rudaceous material accumulated in the base of the Kovernino Formation. Rare occurrence of coarse material in upper parts of the crater lake sequence indicates that some snouts rose above the water level for a long time; some of them (at the west and southwest) existed during the whole Kovernino time.

During the later Bathonian, epeirogenic movements appear in Moscow syneclise. For the Kovernino depression, they cause a significant shallowing of the crater lake. Both central uplift and crater rim have arisen over lake flowage line; as a result, a partial erosion of Kovernino clays has appeared there. Some evidences of removal of Kovernino Formation are recorded in a number of boreholes (51, 179, 181, 302 etc.) drilled in both crater rim and central uplift. Within local lows, a thin cover of sandy sediments accumulated.

In the beginning of the Callovian age, an epicontinental sea has invaded from the north and inundated the whole of the region. The marine environment dominated until the end of the Aptian, but it was unstable: frequent regressions are manifested by diastemes. The most long-term gap in sedimentation occurred in the late Jurassic, when Tithonian deposits were removed in places. However, the denudation was insignificant during that period. When the Middle Jurassic transgression coming, the crater was not fully filled with sediments; so, the maximum thickness of Late Mesozoic deposits is observed in the axial part of the annular trough indicating that the submergence of the latter continued until the Barremian at least.

The maximum of Middle Jurassic-Cretaceous transgression was in the Hauterivian and Barremian. However, this epicontinental sea never inundated the central uplift (including the central pit). All Middle Jurassic and Cretaceous formations are terrigenous and represented by shallow facies.

During Late Cretaceous, Paleogene, and Early Neogene, no sedimentation processes occurred in the area considered. Late Jurassic and Early Cretaceous deposits were almost completely removed beyond the crater; their poor remnants preserve there only in lows of preceding relief. The most complete sequences remain whole in the crater depression alone. In addition, variegated breccias at the ring terrace as well as rare snouts of the central uplift undergone erosion.

The present-day relief of the area started to form at the end of the Miocene when brackish water has reached there following the ancient Volga valley (Pra-Volga). It dammed up streams of the Volga basin, and a barrier lake appeared east of the Volga. This lake was filled in by a thick sequence (more than 40 m) of sandy sediments, which, at last, covered entirely the central uplift. A subsequent regression significantly lowered the basis of the streams of Volga basin. All river and ravine valleys were again deeply cut into Miocene loose deposits; the most of the latter were obliterated.

In the Early Quaternary, an alternation of glacial and interglacial periods occurred. During glaciations, thick ice sheets flattened the surface, formed morainic deposits of various thickness, various fluvioglacial and related alluvial, lacustrine, eolian, and solifluction deposits. During interglacial stages, the deepening of river valleys occurred due to the lowering of the erosion basis. The thickness of glacial till (Kriushino Formation) of the maximum Dnieprian glaciacion peaks 45 m (Kirkov et al. 2019). The present-day morainal topography formed completely due to the latest Moscow glaciation. It was somewhat broken by the formation of the current Volga valley and its tributaries during the Late Quaternary epoch.

References

Blom GI (ed) (1957) Geological map of the USSR on 1:200,000 scale. Sheet O-38-XXVII. Leningrad, VSEGEI, Russian

Bogorodskaya OA, Tumanov RR (1980) Geological map of the USSR on 1:200,000 scale. Sheet O-38-XXVI: Explanatory notes. Russian, Moscow, p 132 (in Russian)

Cohen KM, Finney SC, Gibbard PL, Fan J-X (2013) The ICS international chronostratigraphic chart. Episodes 36:199–204

Fadeeva LI (ed) (2000) Explanatory notes to state geological map of Russian Federation of 1:000 000 scale (new generation), Sheet O-37, (38)—Nizhny Novgorod. Russian, VSEGEI Press, St. Petersburg, p 262 (in Russian)

Holm-Alwmark S, Alwmark C, Ferrière L, Lindström S, Meier MMM, Scherstén A, Herrmann M, Masaitis VL, Mashchak MS, Naumov MV, Jourdan F (2019) An early Jurassic age for the Puchezh-Katunki impact structure (Russia) based on $^{40}Ar/^{39}Ar$ data and palynological investigation. Meteorit Planet Sci 54:1–17. https://doi.org/10.1111/maps.13309

Kirikov VP (ed) (2002) Legend of the state geological map of Russian Federation of 1:1000 000 (3d generation). Central European series of sheets. SIC "Geocentre", Moscow (In Russian)

Kirkov IG, Kochergina VA, Khaidarova DA, Naumov MV (2019) Geological map of Russian Federation on 1:200,000 scale, 2nd generation. Sheet O-38-XXVI (Puchezh): Explanatory notes. Moscow, p 130 (in Press) (in Russian)

Kochergina VA, Kirkov IG, Khaidarova DA (2018) Geological map of Russian Federation on 1:200,000 scale, 2nd generation. Sheet O-38-XXVII (Semenov): Explanatory notes. Moscow, p 90 (in Russian)

Masaitis VL, Naumov MV (1995) An outline model of hydrothermal circulation in impact craters. Trans USSR Acad Sci Earth Sci Sect 377A(6):22–26

Masaitis VL, Pevzner LA (eds) (1999) Seep drilling in the Puchezh-Katunki impact structure. VSEGEI Press, Russian, Saint Petersburg, p 392

Mitta VV, Alekseev AS, Shik SM (eds) (2012) Unified regional stratigraphic chart for the Jurassic of East European platform. Explanatory note. PIN RAS—VNIGNI, Moscow, p 64 (in Russian)

Nalivkin VD, Yakobson KE (eds) (1985) Geology and main patterns of mineral deposit distribution in the USSR, Vol. 1: Russian platform. Nedra Press, Leningrad, p 356 (in Russian)

Naumov MV (2002) Impact-generated hydrothermal systems: data from Popigai, Kara, and Puchezh-Katunki impact structures. In: Plado J, Pesonen LJ (eds) Meteorite impact structures in precambrian shields impact studies 2. Springer, Berlin, pp 117–171

Naumov MV, Larionov AN, Masaitis VL, Mashchak MS, Bogdanova SV, Presnyakov SL, Lepekhina EN (2015) Isotopic age determination of shocked crystalline rocks from the central part of East European Platform (Vorotilovo Deep Drillhole). Reg Geol Metallogeny 61:79–90 (in Russian)

Petrov BV, Kirikov VP (eds) (2006) Geology and mineral commodities of Russia. Volume 1: Western Russia. VSEGEI Press, Saint Petersburg, p 577 (in Russian)

Sazonova IG, Sazonov NT (1967) Paleogeography of the Russian plate in Jurassic and early cretaceous. Nedra Press, Leningrad, p 261 (in Russian)

Tumanov RR, Bogorodskaya OA (1975) Geological map of the USSR on 1:200,000 scale. Sheet O-38-XXVI.VSEGEI, Leningrad (in Russian)

Zorina SO (2012) Jurassic to Paleogene sedimentary sequences of the eastern Russian platform: tectonic-eustatic analysis, sequence stratigraphy, facies transformations, and mineral commodities. Knizhy Perekrestok Publishing, Moscow, 192 pp (in Russian)

Impact Cratering and Origin of the Puchezh-Katunki Structure

Asteroid Impacts and Their Geological Consequences

Boris A. Ivanov

The chapter reviews available data on the impact cratering in the Solar System. Most of data have been collected during robotic space missions to the Moon and other terrestrial planets. The brief introduction into the modern view on the planetary bombardment helps to present rationales for the impact cratering modeling. Morphology and morphometry of extraterrestrial impact craters, comparable is size with the Puchezh-Katunki structure illustrate that all these craters have clearly visible central mound. Crater's depth-diameter relations on planetary bodies with various surface gravity accelerations demonstrate that the simple-to-complex transitional crater diameter increases with decreasing of the surface gravity. The original depth of the Puchezh-Katunki structure and the presence of the central uplift well fit general trends, observed for impact craters on all planetary bodies of the terrestrial type.

1 General Information About Asteroid Impacts

The Puchezh-Katunki (PK) impact structure is one of thousands of impact craters found on Earth, the Moon and other planetary bodies. To immerse the PK description in a general context of impact cratering studies it is instructive to give a short review of main characteristics of known impact craters. The best introduction to the issue has been published by Melosh (1989). To model a crater, we need to assume the nature of the impactor body ("projectile" or "impactor"), its density and the impact velocity.

Many decades of the Solar System dynamics and evolution resulted in a "mainstream" concept, where most of Earth-striking projectiles arrive from the Main asteroid belt (MB). Asteroids normally rotate around the Sun on "safe" orbits, not crossing orbits of terrestrial planets. However, weak repetitive gravity forces (resonance

B. A. Ivanov (✉)
Institute for Dynamics of Geospheres, Leninsky Prospect 38-1, 117939 Moscow, Russia
e-mail: boris_a_ivanov@mail.ru

V. L Masaitis and M. V Naumov (eds.), *The Puchezh-Katunki Impact Crater*,
Impact Studies, https://doi.org/10.1007/978-3-030-32043-0_6

forces) from giant planets like Jupiter and Saturn are able slowly change orbits of asteroids, and an asteroid orbit may become enough elliptical to cross orbits of terrestrial planets (Mercury. Venus, the Earth, and Mars). Collisions between asteroids also can slightly change asteroid orbits. Simultaneously, these collisions produce a number of fragments. Often these fragments are observed as asteroid families. Also, collisions slowly change the size-frequency distribution of small bodies in the MB. We can name these small bodies as "asteroid's fragments", or simply "small asteroids".

The orbital motion of small asteroids and their rotation, in addition to gravity forces, may be changed with weak thermal radiation forces, generated by solar heating/cooling cycles. These effects (Yarkovsky effect, and Yarkovsky–O'Keefe–Radzievskii–Paddack effect, or YORP effect) depend on the body mass, and make small asteroid's orbits more variable than for large asteroids. Gravity resonances and thermal radiation effects may act in chain—thermal radiation forces push a small body to the phase space position where resonances with giant planets act most effectively. The net effect is a permanent weak flux of small bodies from "safe" position to orbits, crossing planetary orbits, making collisions possible.

Being at planetary crossing orbits, small asteroids have relatively frequent flyby in locally strong planetary gravity fields. Close approaches occur much more often than planetary collision. Close approaches change planetary crossing orbits of small asteroids so that a single body cannot occupy the orbit infinitely. After ~3 to ~30 Myr a planetary crosser would fall to the Sun or will be accelerated to orbits beyond Jupiter. Hence, planetary-crossing asteroids observed now have low chances to impact the Earth, other asteroids would be delivered to similar orbits in a future.

It is less probable that a comet nuclear strike the Earth, as Jupiter-family comets originated from the Kuiper belt are observed at planetary-crossing orbits much less frequently than asteroids. Taking simple estimates of impact probabilities for known objects with diameters >1 km (absolute magnitude $H < 18$) we see that only ~5% of these objects are on cometary-like orbits (Werner and Ivanov 2015). Parabolic and long periodic comets seems to have even smaller input to planetary impact bombardment during the last ~3 Gyr—see the review by Dones et al. (2015).

Taking into account the most probable asteroid impact for the primary modeling, we use the known probability distribution to estimate impact velocity. Known $H < 18$ terrestrial crossers occupy orbits giving the average impact velocity of 19–20 km s^{-1} (Werner and Ivanov 2015).

Density of asteroids is known from their fragments, falling to Earth as meteorites (Ostrowski and Bryson 2019). Space missions last decades delivered direct information about the bulk density of a set of asteroids. These estimates show that many asteroids have an appreciable porosity due to fracturing during their collisions with other asteroids (Scheeres et al. 2015). Most probable projectiles in the Earth/Moon system are S-type and C-type asteroids, with bulk densities of 2000–3500 kg m^{-3}. In the first approximation the reasonable initial modeling may be done assuming that the projectile is made of granite-like or basalt-like rocks.

2 Brief Introduction in Crater Mechanics Terminology

Briefly describing the impact cratering processes, we can conditionally split cratering in four distinct stages, differing in time length and in governing physical and mechanical processes. Following Melosh (1989), it is suitable to outline:

Stage 1—the contact and penetration phase. The projectile passes through the atmosphere and impacts the solid (or liquid) surface. At the contact point the shock wave is generated. The maximum pressure here is comparable with the projectile size.

Stage 2—shock wave propagation and the transient cavity growth. Shock wave propagates out of the contact zone, the shock front pressure decay with the propagation distance due to non-isoentropic compression/decompression (energy converts into residual heat) and due to geometrical divergence. After the shock wave passage material keep some residual velocity field out of the impact point. Rarefaction acoustic waves from the free target surface form so-called Z-model flow field, and roughly hemispheric or parabolic cavity starts to grow due to material displacement and ejection to the upper space. This cavity is named the "transient cavity", as only in metals the shape of the final crater is close to the transient cavity shape at the end of its evolution.

Stage 3—the transient cavity modification. The size of the transient cavity is controlled by two main target parameters—gravity field and rock strength. The target material, deforming around the growing transient cavity, experiences the strength resistance, and convert part of its initial kinetic energy into the mechanical "plastic" work. Simultaneously, the flow field, opening the transient cavity by the material displacement uplift the material up in the planetary gravity field, converting the kinetic energy into the potential energy. Part of the flow field energy is lost as the ejecta kinetic energy. The balance between the total "plastic" work against rock strength, and the total potential energy define the so-called cratering regime: if plastic work is the main energy sink, we define that the transient crater is formed in a "strength-dominated" regime. In the alternative case, when the most of the initial flow field kinetic energy is converted into potential energy, we define, that the transient crater is formed in a "gravity-dominated" regime. In a widespread slang we name these cases as "gravity" and "strength" craters.

In dependence on the cratering regime type, a transient cavity, when it reaches the maximum volume, modifies the shape in various ways. First, in strong ductile materials like metals or plasticine, the transient cavity is simply "frozen" with strength forces. Second, in brittle materials, where fragmented rocks behave as a dry sand with the internal friction, a part of cratering walls may slide down to the crater bottom, forming so-called "breccia lens". And third, the residual strength of material around the transient cavity may be low enough to start the collapse of the cavity when the crater floor begins to move up, like the central jet in a water cavity after a pebble dropped to a pond. This case is discussed later in the Section "Computation and Geological Data".

Here we need to state that on all planetary surfaces we see two distinct types of craters—craters, modified mostly by wall slopes are named "simple" craters, and craters, where we see the transient crater uplift, are named "complex" craters.

The floor uplift in complex craters may be hidden under the breccia lens. In this case only depth/diameter ratio comparison may reveal the started floor uplift. We can define these craters as "transitional" from simple to complex. For larger craters the central uplift may be visible as a central peak or peak ring. Largest impact basins have a multiring structure.

Stage 4—the post-impact crater evolution is mostly a set of geologic processes changing the shape of a just-formed impact crater. For large craters these processes include impact melt motion before a final solidification, heat conduction and hot fluids convection to establish a new thermal regime around the new crater. On other planets the post-impact evolution includes as an important part the ongoing bombardment by smaller objects. And, of cause, large craters, as adjacent areas, are modified with "normal" endogenic process like tectonics, volcanism, denudation, and sedimentation.

3 Large Impact Craters on Earth and Other Planetary Bodies

Space missions delivered images of surface on Mercury, Venus, the Moon, Mars, largest asteroids Ceres and Vesta, a dozen of small asteroids. For beyond the asteroid belt objects images are taken for Pluto/Chiron, large satellites of giant planets, for a few of comet nuclei, and, recently, for one of Kuiper belt objects. With rare exception (i.e. Io, Jupiter satellite with the global volcanic activity), all planetary surfaces are covered with impact craters, demonstrating that impact cratering is one of the main processes, formed and renovating solid surfaces of planets and their satellites. Many extraterrestrial planetary bodies have endogenic geologic processes less active than on Earth, or stopped to operate billions years ago. As the result, craters on other planetary bodies often show more pristine impact crater morphology than we see on Earth.

The Moon is studied with multiple space missions, including astronaut's landing and sample return. The absence of atmosphere allows even dust particles to collide the lunar surface without any deceleration. For this reason, we observe lunar impact craters in wide range of sizes—from micron-size craters from micrometeorite impacts to impact basins with 1000 km diameters.

Figure 1 illustrates crater morphologies for a simple, a transitional and, a complex crater. The simple crater has a bowl shape with a small breccia patch at the floor. The transitional crater has irregular floor, but no visible central feature. The complex crater has a flat floor with the central mound. To outline crater shape differences, crater profiles (North to South) are shown in Fig. 2 in coordinates, scaled to a crater rim diameter. In the ideal case, crater rim should be placed at $x/D = \pm 0.5$. Irregular

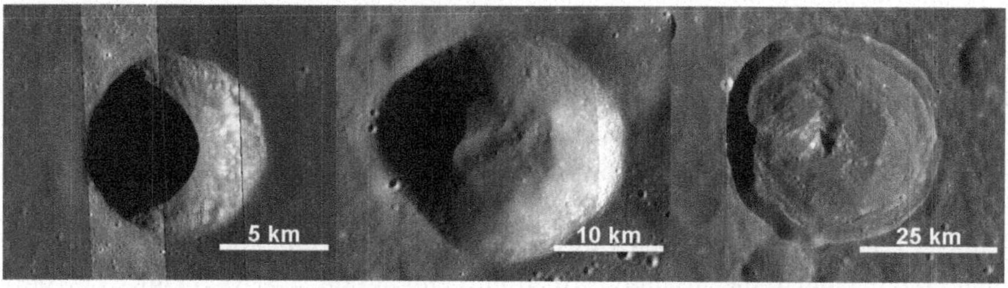

Fig. 1 Typical examples of a simple crater (left, D ~ 8 km, unnamed 359.17E, 8.52S), a transitional crater (center, D ~ 20 km, unnamed, 26.385 E, 39.95 S) and a complex (right, D ~ 40 km 357.859 E, 5.656 S) Hershel crater. Images are generated with the public JMars software (https://jmars.asu.edu/) and based on the LROC WAC mosaic. The left image is enhanced with overlapped LROC images M1188368973LC and M1188368973RC

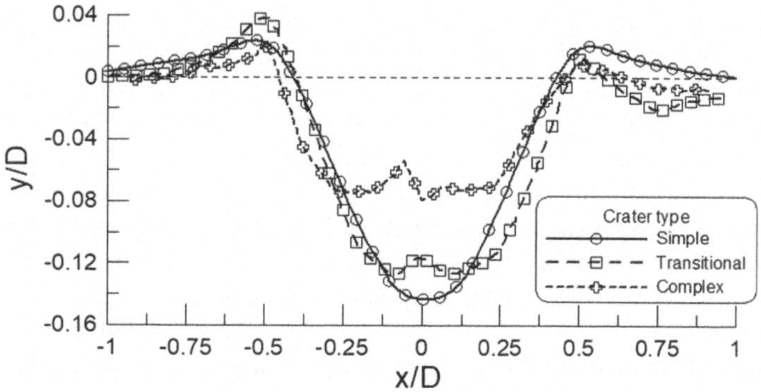

Fig. 2 North-south topographic profiles through three craters, shown in Fig. 4.1. Profiles are constructed with the public JMars software and based on LOLA 1024 ppd (point per degree) topographic map

lunar surface makes hard to find the pre-impact ground level; an approximate position $y = 0$ is shown with a dashed line. The scaled plot well illustrates how a crater depth to diameter ratio, d/D, decreases at the transition from simple to complex craters.

A set of papers presents the statistics of d/D variation with size. First Apollo-era measurements of the lunar crater depth revealed that simple craters are self-similar in a shape, and their d/D ratio is close to 0.2 in a wide crater diameter range from 100 m to 15 km (Pike 1977). The constant d/D means that the crater depth increases linearly with diameter:

$$d \approx 0.2D \tag{1}$$

This is close to crater depth in laboratory high-velocity impact experiments with dry sand, and this fact additionally approves the term "simple" craters (Schmidt and Housen 1987). One should note, that small lunar craters ($D < 1$ to 2 km) degrade faster than larger craters (Ivanov 2018), and below D ~ 1 km we observe shallow degraded craters with depths below Eq. 1 (Stopar et al. 2017). Here Eq. 1 is valid only for fresh craters.

In a gravity cratering regime (see Section "Brief Introduction in Crater Mechanics Terminology"), crater rim profiles should be similar in shape—all measurable values like depth and rim height should be linearly proportional to the crater rim diameter (Housen et al. 1983). However, there is no final opinion at which diameter lunar craters change the strength formation regime to the gravity regime. The only one paper by Moore et al. (1974) mentioned explicitly D ~ 300 m as the effective strength/gravity transition diameter. However, the problem seems to be connected to a non-uniform character of the lunar surface. The upper layer is a regolith, underlined with fragmented rock/soil mixture with rock fragment size and percentage increasing with depth (Heiken et al. 1991). Cohesion and dry friction could vary with depth. Intuitively it seems reasonable to assume that lunar rocks become more competent with depth. If so, there is not a single strength/gravity crater regime transition. It is a matter of a future study to understand how strength and gravity interplay in a wide range of crater diameters. A good introduction in the problem is the numerical modeling study of the strength/gravity transition in uniform targets by Prieur et al. (2017).

The simplest quantitative characteristic of the simple/complex crater transition is given with measurements of the crater depth, d, growth with the crater diameter, D. For the Moon the function d(D) has been measured in 1970s, and these measurements continue now with better elevation measurements at the lunar surface. First systematic measurements are published by Pike (1977), and recent papers sufficiently improve Pike's data (Kalynn et al. 2013; Stopar et al. 2017; Krüger et al. 2018; Osinski et al. 2019). Figure 3 illustrates the current understanding of the lunar d(D) function.

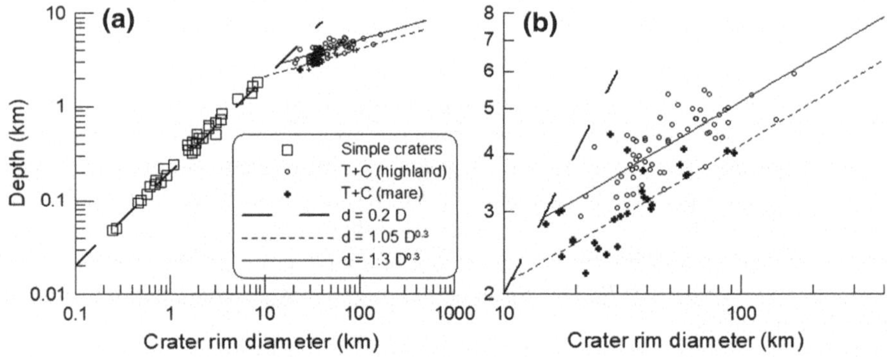

Fig. 3 Rim-to-floor lunar crater depth as a function of the crater diameter. Data for simple craters are from (Stopar et al. 2017). Data for transitional (T) and complex (C) craters are from (Kalynn et al. 2013). Approximated exponent relations (solid, dashed, and dotted curves) are plotted to give a guide to the data cloud and not correspond to statistical method. For power law regression see primary publications. (**a**) The whole crater diameter range, (**b**) The blowup for complex craters

A reader should take into account relatively large uncertainties for large crater measurements. The roughness of the lunar surface and crater rim crest irregularity results in individual depth values from 0.1 to 2 km—see, for example, Fig. 3 in Kalynn et al. (2013). Without any first principle theory, we prefer to illustrate the general trends in d(D) for transitional and complex craters with simple power laws, based on primary vision of Pike (1977). Figure 3 illustrates general trends in d(D) with simple relations (d and D are measured in km):

for lunar mare

$$d = 1.05D^{0.3} \tag{2a}$$

for lunar highlands

$$d = 1.3\ D^{0.3} \tag{2b}$$

These simple power relations cross the d(D) dependence (1) at D = 10.7 km on mare, and D = 14.5 km on highlands. We can take these values as a primary estimate for the simple/complex crater transition. The exhaustive review about d(D) approximations for all cratered planetary bodies is published by Robbins et al. (2018).

Looking at the crater morphology, we see a more complex pattern—transitional craters have no visible central peak, but have lower d/D in comparison with a simple crater dependence (1). So morphologic transition to fully developed central peak craters is extended in the D range from ~15 to ~30 km (Krüger et al. 2018).

The reason to have slightly different d(D) trends for mare and highland craters seems to result from local differences in strength parameters in different locations (Osinski et al. 2019). Lunar highlands present an ancient lunar crust, significantly fragmented by the continuous impact bombardment (megaregolith). Lunar mare are multiple basaltic flows, deposited over the same megaregolith. Osinski et al. (2019) believe that layered basaltic flow structures on mare result in relatively small target rock cohesion, reflected in smaller d(D) for complex craters. The problem needs more quantitative study in a future.

On other planetary bodies we observe similar d(D) relations, complicated with local differences between surface composition, gravity acceleration and presence of an atmosphere. In general, the behavior of d(D) is the same as on the Moon—approximately similar dependence for simple craters as (1), and less steep d growth for complex craters (2018). Within goals of this chapter we outline the main feature of the interplanetary d(D) relations, namely—the increase of an effective crater diameter of simple/complex crater, D_{sc}, at bodies with smaller gravity acceleration at the surface.

Venus has the surface gravity acceleration, close to the terrestrial one (8.07 vs. 9.81 ms^{-2}) but much denser atmosphere and higher surface temperature (Ivanov et al. 1986, 1992; Phillips et al. 1991, 1992). The dense atmosphere plays a two-fold role: (1) meteoroid's breakup at the atmospheric entry disrupts high-velocity projectiles, and (2) dense atmosphere could affect the process of ejecta from the transient cavity. As the result we do not find simple deep craters as on the Moon—a single small body cannot reach the surface, and small craters on Venus are presented with crater clusters. In many cases a body could start fragmentation and fragment separation at low altitudes, resulting in a crater formation by a swarm of fragments (Cochrane and Ghail 2006). Consequently, we have no simple craters to compare them directly with simple lunar craters.

The comparison of Venusian and terrestrial large craters is very important due to a similar value of the surface gravity acceleration. However we should compare these craters with some caution, as the dense Venusian atmosphere could change the ejecta deposition pattern near a crater rim (Ivanov et al. 1992; Schultz 1992; Ivanov 2008).

Surface images of Venus are available after synthetic-aperture radar (SAR) missions Venus 15/16 and Magellan. Dense Venusian atmosphere is opaque for visible light, so radar reflections are processed to produce images as variations of reflected signals mainly due to surface roughness and dielectric properties of surface rocks. Morphology of craters, comparable in size with Puchezh-Katunki is exampled in Fig. 4.

Impact craters found on Venus have been measured with radar image distortion techniques (Ivanov 1989; Sharpton 1994), with simple altimetric signals (Ivanov and Ford 1993), and later, after detailed processing of radar images with different incidence angle, with constructed digital terrain models (DTM)—see Herrick et al. (2012). Figure 5 presents the best available crater depth/diameter data for Venus in comparison with terrestrial craters.

Figure 4 illustrates that on Venus, a rocky planet with the gravity, close to the terrestrial one, craters with D ~ 40 km have the same morphological type as the Puchezh-Katunki impact structure. Venusian craters have a flat floor, central mounds, and terraced inner crater slopes. Despite the dense atmosphere, Venusian craters have a clear visible zone of ejecta depositions beyond the crater rim. Due to high target surface temperature (737 K, i.e. 462 °C) high-velocity impacts produce more impact melt than it would be on Earth. More important is that the impact melt on Venus is more mobile due to slower cooling by hotter rock fragments (Ivanov et al. 1992). This longer melt mobility could explains the smooth (radar-dark) appearance of Venusian crater floors.

The depth trend in complex crater depths both for Venus and for Earth (Fig. 5) may be presented with the lunar-like power law

$$d = 0.3D^{0.3} \tag{3}$$

giving at D = 40 km the crater depth of 0.9 km, what is much less than for lunar craters of the same diameter (3–4 km). This difference illustrates how gravity forces control the transient cavity collapse during a complex impact crater formation. The closeness of Venusian and terrestrial d(D) trends illustrates a weak dependence on a water presence. The lower depth of terrestrial craters in sedimentary targets could indicate as the lower material strength, as well as a participation of water in the strength lowering.

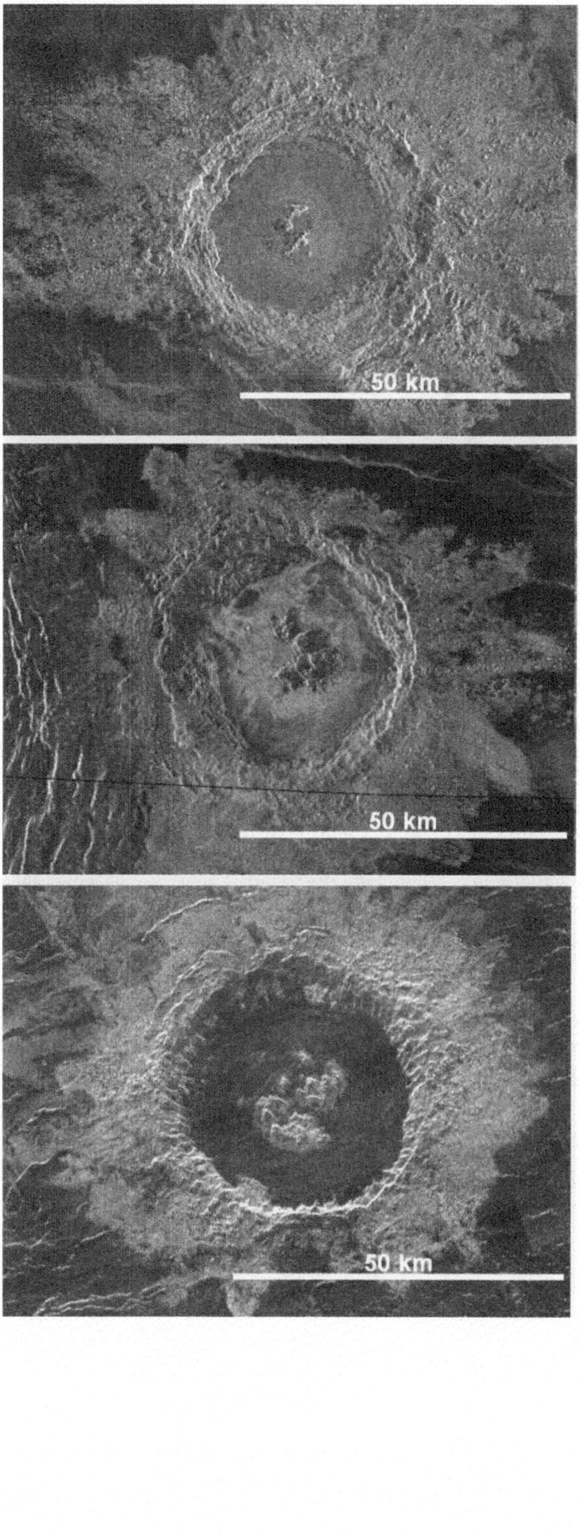
50 km
50 km
50 km

◂**Fig. 4** Three Venusian craters with diameters close to the Puchezh-Katunki impact structure: Xiao Hong (left, D ≈ 38.7 km), La Fayette (middle, D ≈ 39.6 km) and Duncan (right, D ≈ 40.3 km). The visible brightness of the central mound, inner crater walls and continuous ejecta zone are mainly resulted from the enhanced surface roughness, contrasted with smooth (radar-dark) plains outside craters and impact melt pond surface inside. Magellan mosaic F-BIDRs, (Full resolution Basic Image Data Records ~ 75 m/pixel). Publisher: USGS Astrogeology Science Center. Images are generated with JMars software (https://jmars.asu.edu/)

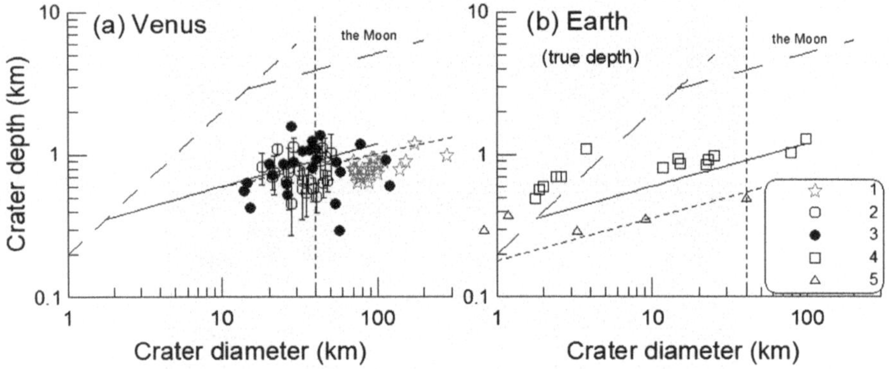

Fig. 5 Available crater depth estimates for Venus (**a**) and Earth (**b**). 1—direct altimeter measurements (Ivanov and Ford 1993), 2—DEM (Neish et al. 2017), 3—DEM (Cochrane and Ghail 2006). Terrestrial data (**b**) are illustrated with "true" crater depths collected by Grieve et al. (1981) for craters in crystalline (4) and sedimentary rock targets. Note that data points (2, 3) include rim heights into rim-to-floor depth, while data points (1) on Venus and (3, 4) on Earth presents so called "terrain-to-floor" depth, i.e. the crater floor depth below an estimated pre-impact surface level. Long-dashed curves (lines in the log-log plot) present lunar data (Fig. 4.3). Solid curves trough data clouds are not fitting lines and are drown with the same power law as on Fig. 4.3 (d ~ $D^{0.3}$) to fit the data at D ~ 40 km to compare later the "depth deficit" versus a hypothetical simple crater on different planetary bodies. The vertical dotted line shows approximately the Puchezh-Katunki diameter

Mercury and Mars. Another pair of planetary bodies with close surface gravity accelerations, but different near-surface geology is Mercury ($g = 3.7$ m s^{-2}) and Mars ($g = 3.72$ m s^{-2}). Morphology of simple craters here changes with a crater diameter in the same way as on the moon—from simple craters to complex craters with central uplifts (Fig. 6).

For the comparison we plot on Fig. 6 lunar-like curves (lines in log-log coordinates) for complex crater's d(D):

$$\mathrm{D} = 0.75 * \mathrm{D}^{0.3} \tag{4}$$

for Mercury and

$$\mathrm{D} = 0.8 * D^{0.3} \tag{5}$$

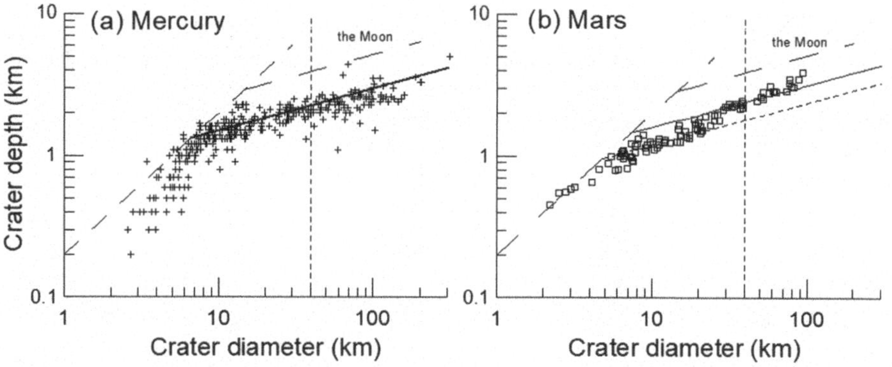

Fig. 6 Available data on the depth/diameter, d(D), relations for impact craters measured on Mercury (**a**) and Mars (**b**). Long dashed curves (lines in log-log coordinates show lunar d(D) trends for simple and complex craters. Data for Mercury are collected by Susorney et al. (2016), and for Mars by (Tornabene et al. 2018). The vertical dotted line shows approximately the Puchezh-Katunki diameter

for Mars.

Here we do not want to discuss the visible deflection of d(D) from Eq. 5, and simply plot the curve (5) to cross measured data around D = 40 km.

Images of two impact craters on Mercury and on Mars with diameters near 40 km are shown in Fig. 7.

Craters with sizes, close to the Puchezh-Katunki impact structure, on all five planetary bodies have similar morphology: relatively flat floor with the central mound. The shape of central mounds slightly varies even at the same planetary body, exampled with Venusian craters in Fig. 4. It seems to reflect local difference in target

Fig. 7 Complex impact craters with diameters 38 km on Mars (left) and on Mercury (right). Mars coordinates are 118.2 E, 39.5 S, HRSC image H8728_0000_ND3. Mercury coordinates are 122.03 E, 10.0 S, MDIS image EN1016060659 M. Images are generated with JMars software (https://jmars.asu.edu/)

Table 1 The depth estimates for a crater with D = 40 km on terrestrial planetary bodies

Planetary body	Gravity acceleration, g, m s^{-2}	Crossing diameter[a] D* (km)	Crater depth (km)
Moon mare highlands	1.62	10.7 14.5	3.2 3.9
Mars	3.71	7.2	2.4
Mercury	3.7	6.6	2.3
Venus	8.9	1.8	0.9
Puchezh-Katunki (Earth)	9.81	–	0.8–1.5

[a]The diameter where d(D) trends for complex craters cross the trend for simple craters (see Figs. 3, 5, and 6)

properties, but some researches connect this difference with a possible effects of different impact velocities (Osinski et al. 2019).

Crater depths for complex craters (in most cases—the depth of a flat floor around the central mound) systematically vary with the crater diameter, growing much slower than d(D) for simple craters. Theoretically the value d/D for complex craters depends on the ratio D/D_{sc}, where D_{sc} is the simple-to-complex transitional diameter (Croft 1985; Wünnemann and Ivanov 2003). The value of D_{sc} increases with the decrease of the gravity acceleration, e.g. Pike (1980) suggested that $D_{sc} \sim g^{-1}$. Recent data, including impact craters on icy bodies, demonstrates a possibility of a slightly less strong dependence on g (Hiesinger et al. 2016), and this issue still demands more analysis. For a practical comparison with the Puchezh-Katunki structure we use approximations with the lunar-like d(D) slope to trace how the depth for D = 40 km crater varies from the Moon to Venus (Table 1). As the gravity acceleration raises up, the diameter of an effective transition from simple to complex crater's d/D decreases from $D_* \sim 15$ km to $D_* \sim 2$ km, and the crater depth decreases from 3 to 4 km on the Moon to ~1 km on Venus.

The presence of the central uplift in the Puchezh-Katunki structure and its estimated pristine depth about 0.8 to 1.5 km (with the uncertainty where was the crater floor just after impact) make the structure a typical impact crater of its size on terrestrial planets in our Solar System.

References

Cochrane CG, Ghail RC (2006) Topographic constraints on impact crater morphology on Venus from high-resolution stereo synthetic aperture radar digital elevation models. J Geophys Res Planets 111:E04007. https://doi.org/10.1029/2005JE002570

Croft SK (1985) The scaling of complex craters. J Geophys Res 90:C828–C842

Dones L, Brasser R, Kaib N, Rickman H (2015) Origin and evolution of the cometary reservoirs. Space Sci Rev 197:191–269

Grieve RAF, Robertson PB, Dence MR, Schultz PH (1981) Constraints on the formation of ring impact structures, based on terrestrial data. In: Merill RB (ed) Multi-ring basins: formation and evolution. Pergamon Press, New York and Oxford, pp 37–57

Heiken GH, Vaniman DT, French BM (1991) Lunar sourcebook—a user's guide to the moon. Cambridge University Press, Cambridge, UK, p 753

Herrick RR, Stahlke DL, Sharpton VL (2012) Fine-scale venusian topography from magellan stereo data. Eos, Trans Am Geophys Union 93(12):125–126

Hiesinger H, Marchi S, Schmedemann N, Schenk P, Pasckert JH, Neesemann A, O'Brien DP, Kneissl T, Ermakov AI, Fu RR, Bland MT, Nathues A, Platz T, Williams DA, Jaumann R, Castillo-Rogez JC, Ruesch O, Schmidt B, Park RS, Preusker F, Buczkowski DL, Russell CT, Raymond CA (2016) Cratering on Ceres: implications for its crust and evolution. Science 353(6303):aaf4759

Housen KR, Schmidt RM, Holsapple KA (1983) Crater ejecta scaling laws—fundamental forms based on dimensional analysis. J Geophys Res 88:2485–2499

Ivanov BA (1989) The morphometry of impact craters on Venus. Astronomicheskii Vestnik 23:39–49 (in Russian)

Ivanov B (2008) Impact crater ing on venus: ballistic hole in the atmosphere (abs.). In: Proceedings European planetary science congress 2008, Münster, Germany, p 341

Ivanov BA (2018) Size-frequency distribution of small lunar craters: widening with degradation and crater lifetime. Sol Syst Res 52(1):1–25

Ivanov BA, Ford PG (1993) The depths of the largest impact craters on Venus. Lunar and Planetary Science XXIV. Houston, TX, pp 689–690

Ivanov BA, Bazilevsky AT, Krivchkov VP, Chernaia IM (1986) Impact craters of venus—analysis of venera 15 and 16 data. J Geophys Res 91:D413–D430

Ivanov BA, Nemchinov IV, Svetsov VA, Provalov AA, Khazins VM, Phillips RJ (1992) Impact cratering on Venus: Physical and mechanical models. J Geophys Res 97(E10):16167–16181

Kalynn J, Johnson CL, Osinski GR, Barnouin O (2013) Topographic characterization of lunar complex craters. Geophys Res Lett 40:38–42

Krüger T, Hergarten S, Kenkmann T (2018) Deriving morphometric parameters and the simple-to-complex transition diameter from a high-resolution, global database of fresh lunar impact craters ($D \geq \sim 3$ km). J Geophys Res Planets 123:2667–2690

Melosh HJ (1989) Impact cratering—a geologic process. Oxford University Press, Oxford—New York, p 245

Moore HJ, Hodges CA, Scott DH (1974) Multiringed basins—illustrated by orientale and associated features. In: Proceedings 5th lunar and planetary science conference, Pergamon Press, New York, pp 71–100

Neish CD, Herrick RR, Zanetti M, Smith D (2017) The role of pre-impact topography in impact melt emplacement on terrestrial planets. Icarus 297:240–251

Osinski GR, Silber EA, Clayton J, Grieve RAF, Hansen K, Johnson CL, Kalynn J, Tornabene LL (2019) Transitional impact craters on the moon: insight into the effect of target lithology on the impact cratering process. Meteorit Planet Sci 54:573–591

Ostrowski D, Bryson K (2019) The physical properties of meteorites. Planet Space Sci 165:148–178

Phillips RJ, Arvidson RE, Boyce JM, Campbell DB, Guest JE, Schaber GG, Soderblom LA (1991) Impact craters on venus—initial analysis from Magellan. Science 252:288–297

Phillips RJ, Raubertas RF, Arvidson RE, Sarkar IC, Herrick RR, Izenberg N, Grimm RE (1992) Impact craters and venus resurfacing history. J Geophys Res 97:15923–15948

Pike RJ (1977) Size-dependence in the shape of fresh impact craters on the moon. In: Roddy DJ, Pepin RO, Merrill RB (eds) Impact and Explosion Cratering: Planetary and Terrestrial Implications. Pergamon Press, Oxford, pp 489–509

Pike RJ (1980) Control of crater morphology by gravity and target type—Mars, Earth, Moon. In: Proceedings of the eleventh Lunar and Planetary Science Conference, pp 2159–2189

Prieur NC, Rolf T, Luther R, Wünnemann K, Xiao Z, Werner SC (2017) The effect of target properties on transient crater scaling for simple craters. J Geophys Res Planets 122(8):2017JE005283

Robbins SJ, Watters WA, Chappelow JE, Bray VJ, Daubar IJ, Craddock RA, Beyer RA, Landis M, Ostrach LR, Tornabene LL, Riggs JD, Weave BP (2018) Measuring impact crater depth throughout the solar system. Meteorit Planet Sci 53:583–637

Scheeres DJ, Britt D, Carry B, Holsapple KA (2015) Asteroid Interiors and Morphology. In: Michel P, DeMeo FE, Bottke FW (eds) Asteroids IV. University of Arizona Press, Tucson, pp 745–766

Schmidt RM, Housen KR (1987) Some recent advances in the scaling of impact and explosion cratering. Int J Impact Eng 5:543–560

Schultz PH (1992) Atmospheric effects on ejecta emplacement and crater formation on venus from Magellan. J Geophys Res 97(E10):16183–16248

Sharpton VL (1994) Evidence from Magellan for unexpectedly deep complex craters on Venus. In: Dressier BO, Grieve RAF, Sharpton VL (eds) Large Meteorite Impacts and Planetary Evolution. Geological Society of America, Boulder, CO, pp 19–27

Stopar JD, Robinson MS, Barnouin OS, McEwen AS, Speyerer EJ, Henriksen MR, Sutton SS (2017) Relative depths of simple craters and the nature of the lunar regolith. Icarus. https://doi.org/10.1016/j.icarus.2017.05.022

Susorney HCM, Barnouin OS, Ernst CM, Johnson CL (2016) Morphometry of impact craters on Mercury from MESSENGER altimetry and imaging. Icarus 271:180–193

Tornabene LL, Watters WA, Osinski GR, Boyce JM, Harrison TN, Ling V, McEwen AS (2018) A depth versus diameter scaling relationship for the best-preserved melt-bearing complex craters on Mars. Icarus 299:68–83

Werner SC, Ivanov BA (2015) Exogenic dynamics, cratering, and surface ages (chapter 10.10). In: Schubert G (ed) Treatise on geophysics, 2nd edn. Elseiver, Oxford, pp 327–365

Wünnemann K, Ivanov BA (2003) Numerical modelling of the impact crater depth–diameter dependence in an acoustically fluidized target. Planet Space Sci 51(13):831–845

Modeling of the Impact Structure Origin

Boris A. Ivanov

The numerical model of the Puchezh-Katunki formation processes is described. The model results are preceded with a short description of all model components: the concept of hydrocodes, rock equation of states, brittle/ductile rock damaging, and the acoustic fluidization (AF) model, necessary to simulate the temporary dry friction reduction in rocks around forming impact craters. The computer model of the Puchezh-Katunki crater presents a set of variants, simulated the vertical impact of a rocky projectile (asteroid) 3.5–4 km in diameter with the velocity 11–16 km s^{-1}, resulted in a model crater shape, resembling the observed structure geometry. The recorded maximum shock pressures in specimens from the VDW recovered core are close to modeled values, supporting the correctness both of geologic interpretations and general principles of the cratering numerical model.

1 Hydrocodes and Equations of State

Any computer model of natural dynamic processes combines various particular mathematical models of the matter motion. The core model uses hydrodynamic equations of media motion under internal and external forces. These equations represent three main conservation laws, namely the mass, momentum, and energy conservation. Many basic problems may be solved analytically. Problems not solving analytically are treated with various approximation techniques. One of main techniques to solve hydrodynamic equations of motion is the finite difference approach. In this approach the limited in size area of interest is separated into discrete elements ("cells"), and the general problem converts to the motion of a cell with forces (pressure and stresses) from neighbor cells during a short time period, named a time step.

B. A. Ivanov (✉)
Institute for Dynamics of Geospheres, Moscow Leninsky Prospect 38-1, 117939, Russia
e-mail: boris_a_ivanov@mail.ru

V. L Masaitis and M. V Naumov (eds.), *The Puchezh-Katunki Impact Crater*,
Impact Studies, https://doi.org/10.1007/978-3-030-32043-0_7

In compressible materials, like rocks and other geomaterials, the finite sound speed allows us to find a minimum value of a time step, to restrict cell interactions by their neighbors only. Computer codes, named *hydrocodes*, solve the problem by innumerous cycles over space and time. The main steps the code need to do are

- Compute pressure in each cell
- Compute accelerations in each cell from cell-to-cell pressure difference
- Compute accelerations in each cell from deviatoric stress differences
- Compute the 'new" volume (and density) in each cell
- Compute the "new" energy in each cell (and pressure and temperature)

As soon as the code computes "new" values, one can make a small-time step forward and compute displacements of material points. The consequent change in cell volumes results in cell compression/decompression (the material density increases/decreases), the pressure in a cell increases/decreases, and the code is ready to make the next time step. This very brief list of operations gives a general understanding of particular models of the material behavior the code uses. To compute pressure from density and temperature (or internal energy) the code use so called *equations of state* (EOS), individual for each material. To compute *strength and friction* influence the code use model descriptions of the brittle and ductile material behavior, based mainly on laboratory tests of geomaterials. Below we briefly list models, used here.

Hydrocode. The solver of hydrodynamic equations of motion in a discretized form nicknamed here as "SALEB" is based on the original hydrocode SALE developed in the Los-Alamos Laboratory, USA (Amsden et al. 1980). The original code treats one material motion in Eulerian and Lagrangian mode. H.J. Melosh with co-authors have advanced SALE to treat multimaterial impact problems with rock equations of state (Watts et al. 1991), and with brittle strength properties (Melosh et al. 1992). The basic code has been improved to be able to compute micromaterial problems in the Eulerian mode in a style proposed for other hydrocodes like CTH (McGlaun et al. 1990) and SOVA (Shuvalov 1999). The SALEB code has been tested for many impact problems (Ivanov et al. 1997a, b, 2010; Ivanov 2005). The new branch of the SALE and SALEB evolution, named iSALE is developed and maintained by research groups at the Museum für Naturkunde, Berlin, and Imperial College, London (https://isale-code.github.io/).

Equation of state (EOS). Equations of state (EOS) is a system of relationships which allow us to describe the thermodynamic state for materials of interest. In its basic form, an EOS should define what is the pressure in the material at a given density and temperature. In extended forms an EOS should define also the phase state of material (melting, vaporization, dissociation, and ionization) as well as all useful derivatives of basic parameters (like sound speed and heat capacity), and transport properties (like heat conductivity etc.). Hence, the full EOS should include a lot of parameters and physical processes. The engineering kind of EOS's typically includes experimentally measured data with a system of equations to interpolate and extrapolate tabulated data (best tables, seems, are published for water—the main active agent for steam engines and turbines).

The usual starting example is the equation of state for the ideal gas—gas of molecules interacting only by collisions. Several equations for the ideal gas below are presented in forms, making a bridge to recognize simple "first principle" terms in more complex EOS discussed below. For an ideal gas pressure, p, specific volume, V, and temperature, T, are related with the equation of state

$$p\,V = (R/m)\,T \tag{1}$$

where $R = 8.317$ J/mole K is the universal gas constant, and m is the molecular weight of gas. The material specific volume is the reciprocal of material density, $\rho = 1/V$, and Eq. 1 may be written as

$$p = (R/m)\,\rho\,T \tag{2}$$

The internal specific energy, e, of the ideal gas is

$$e = RT/(\gamma - 1)\,m \tag{3}$$

and the pressure-density relation may be presented as

$$p = a\,\rho\,e \tag{4}$$

where $a = \gamma - 1$ and γ is the ratio of specific heats of the gas: $\gamma = 5/3$ for mono-atomic gases, 1.4 for diatomic gases and 1.33 for more complex gases.

Equation 4 is ready to be used in numerical simulations. As it was mentioned above, the completed previous time step of a hydrocode gives values of "new" density and specific internal energy, and Eq. 4 defines the "new" pressure to make the next time step. Two important notes: (1) for an ideal gas all the internal energy is the energy of the thermal motion of molecules; (2) the explicit value of temperature is not needed for computations—temperature may be computed from internal energy with Eq. 3.

In contrast to the ideal gas, atoms in solids are tightly bound with inter-atomic forces. These forces have a minimum (zero) value at a specific distance between atoms. If this distance is decreased (in compression) or increased (in extension), inter-atomic forces tend to return atoms into the equilibrium position. With some simplification we can say that at normal condition (ambient pressure and room temperature) the density (so called *normal density*) of solids is defined by the position of atoms, separated at a distance, where inter-atomic forces are zero (or at a level to balance 1 bar atmospheric pressure). Deviation from the normal density, ρ_0, results in pressure change: Pressure increases when $\rho > \rho_0$ (compression), and decreases when $\rho > \rho_0$ (tension).

One can imagine inter-atomic forces as mechanical springs connecting neighbouring atoms. Compression or extension results in spring's loading, creating attractive or repulsive forces between atoms. This analogy is useful to introduce the specific internal energy resulting from the change of distance between atoms. The imaginary "springs" of inter-atomic forces store a potential energy being extended or compressed. Hence, even without a heating, compression/extension of bound atoms and molecules in condensed states (solid or liquid) create the specific internal energy, which should be included into EOS for numerical modelling. From a time of EOS elaboration for metals, this energy is often named "nuclear" energy—energy of compression/extension of a lattice with nucleus in nodes. As these energy and pressure do not depend on the material temperature, they are often nicknamed as "cold" energy and "cold" pressure, following the classic work by Zeldovich and Raizer (1966–67).

For small value of compression (typically measured by density ratio, $\mu = \rho/\rho_0$, or the density ratio minus one, $\eta = \rho/\rho_0 - 1 = \mu - 1$ the increase of pressure may be described as a linear function of density increase:

$$p = K(\rho/\rho_0 - 1) = K\,\eta \tag{5}$$

where K is the bulk compressibility module.

The linear relationship does not work for the large compression of rocks and minerals, as the modulus K increases. The usual form for a static compression is the Murnaghan equation (Murnaghan 1944):

$$p = \frac{K_{s0}}{n}\left(\left(\frac{\rho}{\rho_0}\right)^n - 1\right)^{-n} \tag{6}$$

where K_{S0} is the bulk module at normal density $\rho = \rho_0$ and n is an exponent which controls the rate of K_S growth with compression. At large compressions the $p(\rho)$ behaviour of solids is more complicated, and a variety of analytical expression has been proposed—see textbooks by Poirier (2000) and Anderson (2005). Here we note that for an arbitrary compressional law one finds the specific internal energy ("cold" energy) integrating the $p(\rho)$ function

$$e_c = -\int_{V_0}^{V} p\,dV = \frac{1}{\rho_0}\int_{\rho_0}^{\rho} p\,d\rho \tag{7}$$

At elevated temperatures solids accumulate the energy of atomic vibrations—thermal internal energy—e_{th}, in addition to the "cold" energy. Heating (increase of the thermal energy) creates also an additional pressure (thermal pressure). At constant ambient conditions, the thermal pressure tends to expand a solid body (the usual thermal expansion), to balance the internal and ambient pressure. The thermal expansion increasing the body volume, distances between atoms increase, decreasing the "cold" pressure. The interplay of these processes is expressed via a combination of material parameters, named the Grüneisen parameter G:

$$G = \frac{\alpha K_s}{\rho c_p} \tag{8}$$

where α is the thermal expansion coefficient, and c_p is the heat capacity at constant pressure. Using this coefficient, the "thermal" pressure, p_{th}, in solids may be approximated by an analogy with the ideal gas (4) via the "thermal" energy, e_{th}:

$$p_{th} = G\,\rho\,e_{th} \tag{9}$$

The very useful approximated EOS, named "Mie-Grüneisen" EOS, is constructed by summation of the "cold" part and the "thermal" part. The "cold" pressure depends only on the material density, $p_c = p_c(\rho)$, and the internal "cold" energy is computing as an integral (7). The Grüneisen parameter is also function of density, $G = G(\rho)$. The pressure and internal energy in compressed solids are expressed as

$$p = p_c(\rho) + p_{th}(\rho, T) = p_c(\rho) + G(\rho)\rho e_{th}(T) \tag{10}$$

$$e = e_c(\rho) + e_{th}(T) \tag{11}$$

Most hydrocodes compute pressure and density directly through the internal energy, and the temperature is computed separately to find, for example, melting point.

The expression for $p_c(\rho)$ may be much more complicated than examples shown above (5 and 6). Approximation of experimental data on high pressure compression of rock and minerals has allowed to construct equations of state for all materials, useful for the interpretation of deep planetary interiors (Zharkov and Kalinin 1968; Zharkov and Kalinin 1971; Anderson 2005).

For an approximate numerical modelling of impacts the temperature may be estimated if we know the thermal energy, e_{th}, temperature dependence at a range of ρ and *T*. In a simple approximation of a constant heat capacity coefficient, *c*, we can assume, for example, that

$$e_{th}(\rho, T) = e_{th}(\rho, T_{room}) + c(T - T_{room}) \tag{12}$$

where T_{room} is the "normal" temperature. For many rock-forming minerals the heat capacity *c* at high temperature is close to a constant—according to measurements and the Debye theory being in the range of 1000–1200 J kg^{-1} K^{-1}. One can use more sophisticated approach with a variable *c(T)*. The simple approximation (12) has been recommended for first-order estimates in the impact numerical modelling (Do and Benson 2000; Ivanov et al. 2002).

A specific version of Mie-Grüneisen's EOS was proposed by Tillotson (1962) for metals[1] and expanded later for rocks by (Allen 1967; Melosh 1989). Originally Tillotson's EOS has been constructed for very large compressions typical for nuclear explosions, and a focus of this EOS design to reach correct approximations at highest compressions and temperatures. This dictates a deviation from a standard Mie-Grüneisen EOS form (10). However, the spirit of this approximation is the same.

[1]https://apps.dtic.mil/dtic/tr/fulltext/u2/486711.pdf (available May 5, 2019)

Table 1 Tillotson's EOS parameters used in the model (Eq. 13)

Parameter	Basement ("granite")	Sediments ("wet tuff")
ρ_0, kg m^{-3}	2650	1790
A, GPa	53	10
B, GPa	53	10
a	0.4	0.5
b	0.1	1.3
E_0, J	4.87×10^8	11×10^6

Tillotson's EOS is widely used in numerical models as it includes very approximate, but good enough for qualitative estimates of the shock vapor expansion. Time to time researchers try to improve Tillotson's EOS and to clarify the thermodynamics followed from this EOS (Brundage 2013). Here we illustrate only "compressional" part of Tillotson's EOS, used to estimate shock melting in the model.

In compression ($\rho > \rho_0$) Tillotson's EOS describes how pressure increases with density and internal energy;

$$p = A\mu + B\mu^2 + \rho e\left(a + \frac{b}{\frac{e}{E_o\eta^2} + 1}\right) \tag{13}$$

where five constants (A, B, a, b and E_O) typically result from a fit shock compression experiments, (see μ and η definitions above Eq. 5).

Examples of Tillotson's parameters, used in the modelling presented below, are listed later in Table 1.

To use Eq. 12 for temperature estimates one needs to "separate" the internal energy into "cold" and thermal constituents. Technically, the modelling code has a special subroutine, pre-computed "cold" values of pressure and energy versus density for a discrete set of densities. The procedure use Eq. 13 assuming zero thermal input (e.g., constant temperature). The standard Runge–Kutta fourth-order method is used to integrate Eq. 7 with $p(\rho, e)$ defined with Eq. 13 for $e = e_c$. Results are recorded in two tables $p(\rho)$ and $e_c(\rho)$. Each program call to the equation of state subroutine interpolates these tables to find "cold" energy for a given density, using the inversion of the Eq. 12 to estimate the current local temperature.

More advanced Analytical Equation of Sate (ANEOS) is a lengthy computer program to follow all previous approaches (separation of "cold" and thermal energies and pressures) in a more flexible way (Thompson and Lauson 1972). The original ANEOS code has been improved for the more realistic description of vaporization (Melosh 2007). The standard usage of the ANEOS includes pre-computation of tables, which the hydrocode interpolate to compute pressure, density, temperature and internal energy in each computational cell at each timestep. Tables for many rocks and minerals are included in the iSALE package (for more information please inspect instructions at https://isale-code.github.io/index.html).

The numerical model of the Puchezh-Katunki impact, described later in this chapter uses mostly Tillotson's equation of state and, in several model runs, pre-computed ANEOS-based tables.

Figure 1 illustrates the Hugoniot curve for granite, computed with Tillotson's equation of state with parameters, listed in Table 1, and with ANEOS-based tables, constructed with input parameters published by Pierazzo et al. (1997). Hugoniot curves are computed with the 1D Lagrangian mode of the SALEB code with the rock strength model used in the impact cratering modeling. Hence, data shown in Fig. 1 take into account the heating from mechanical work against strength forces (described in the next subsection).

Strength and friction in the model. Strength properties of solids are usually described with the elastic stresses increment computations from strain increments. The strength model verifies if stresses in the material are below or above the model stress level. Stresses above the strength level are relaxed too keep the stress state at the critical level. Strength brittle/ductile models used in the SALEB hydrocode are described elsewhere (Ivanov et al. 1997b; Collins et al. 2004). Here we briefly repeat main definitions.

In each time step the hydrocode compute displacements of material points during the time step. These displacements allow us to compute accumulated strains,

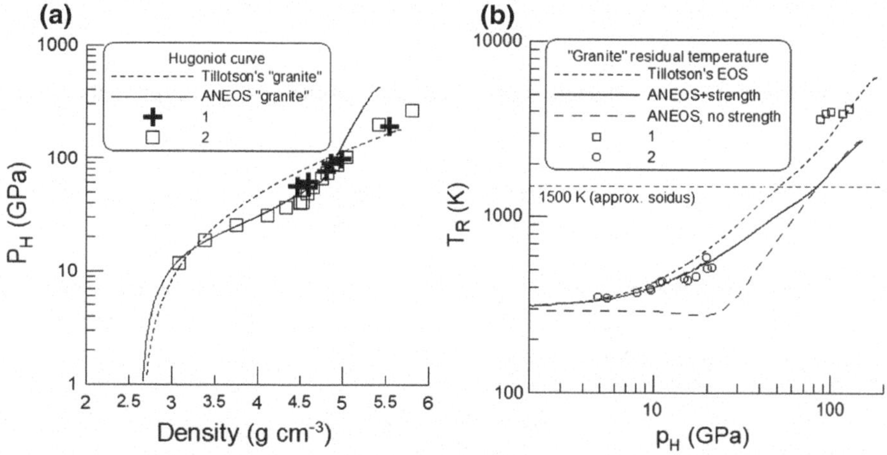

Fig. 1 **a** Hugoniot curves ("shock adiabates") modelled with Tillotson's EOS and with ANEOS tables based on (Pierazzo et al. 1997), presenting the shock front pressure, p_H, as a function of the maximum material density at the shock front. Experimental data for granite (1) and quartz (2) are taken from the data compendium collected by Trunin et al. (2001). **b** Computed residual temperatures of "granite", T_R, after the pressure release to 1 bar as a function of the shock pressure. Experimental data for quartz residual temperatures (1 and 2) are from (Raikes and Ahrens 1979; Boslough 1988). The prominent difference in $T_R(p_H)$ for ANEOS based curves computed with and without strength model is in accord with a recent discussion (Kurosawa and Genda 2018; Melosh and Ivanov 2018)

and the local strain rate. The strain is separated into a volumetric strain (compression/expansion of a material particle), and deviatoric strains (the shape change, like shear a material particle).

In axial symmetry coordinates (radial coordinate x, and vertical coordinate y), used here, the deviatoric strain rate tensor $\boldsymbol{D_e}$ has four non-zero components:

$$e_{ij} = \begin{pmatrix} \frac{\partial u_x}{\partial x} - \frac{1}{3}\varepsilon & \frac{1}{2}\left(\frac{\partial u}{\partial y} + \frac{\partial v}{\partial x}\right) & 0 \\ \frac{1}{2}\left(\frac{\partial u}{\partial y} + \frac{\partial v}{\partial x}\right) & \frac{\partial v}{\partial y} - \frac{1}{3}\varepsilon & 0 \\ 0 & 0 & \frac{u}{x} - \frac{1}{3}\varepsilon \end{pmatrix} \quad (14)$$

where u and v are velocity components along x and y axis correspondingly, and ε is the volume strain rate, defined with the cell volume change during the timestep.

The stress tensor is also separated in a volumetric part (pressure) and a deviatoric part, responsible for shear stresses. For the usual notations of three main stresses σ_1, σ_2, and σ_3, the average stress is defined as the pressure

$$p = (\sigma_1 + \sigma_2 + \sigma_3)/3 \quad (15)$$

Deviatoric stresses are defined as

$$S_{ij} = \sigma_{ij} - p \quad (16)$$

In the axial symmetry the stress state is defined with 4 stress components:

$$S_{ij} = \begin{Bmatrix} S_{xx} & S_{xy} & 0 \\ S_{xy} & S_{yy} & 0 \\ 0 & 0 & S_{th} \end{Bmatrix} \quad (17)$$

The SALEB hydrocode use linear elastic equations to compute the deviatoric stress changes in each material cell during each time step Δt from the time moment ${}^{m}t$ to the next time moment ${}^{m+1}t\left(\Delta t = {}^{m+1}t - {}^{m}t\right)$:

$${}^{n+1}S_{ij} = {}^{n}S_{ij} + 2Ge_{ij}\,\Delta t \quad (18)$$

where G is the shear modulus, and e_{ij} is the strain rate tensor (14).

Values of shear strain are compared with the local strength limit and are decreased if they are out of the elastic conditions. The deformation, accumulated in a material particle when stresses are limited with the strength limit, are recorded for each particle as a "plastic strain", i.e. strain at timesteps where stresses overcome the local strength limit.

Elastic properties of solid materials are described with in a standard way via the Poisson ratio, ν. The bulk compressional modulus, $K(\rho)$, is defined with the equation of state, described above. The shear modulus, G, as well as longitudinal and shear elastic wave velocities, v_p and v_s, are computed with standard elastic relationships

$$G = 1.5\,K\frac{1-2\nu}{1+\nu} \tag{19}$$

$$v_p = \sqrt{\left(\frac{K+\frac{4}{3}G}{\rho}\right)} \tag{20}$$

$$v_s = \sqrt{G/\rho} \tag{21}$$

To describe a material without deviatoric elastic stresses (e.g., the melted material) the hydrocode change the Poisson ratio value to $\nu = 0.5$, and the acoustic wave velocity, v_b, is computed via the bulk compressibility modulus

$$v_b = \sqrt{K/\rho} \tag{22}$$

The rock strength behaviour in the hydrocode is described with the approximate description of the brittle/ductile behaviour—see details in (Collins et al. 2004). The model is based on laboratory rock tests, including triaxial tests. The approximate, but computer-friendly approach is to check the value of the second invariant of stress deviator, J_2. In the SALEB code this value is computed as

$$J_2 = \frac{1}{6}\left[(\sigma_1-\sigma_2)^2 + (\sigma_2-\sigma_3)^2 + (\sigma_3-\sigma_1)^2\right] \tag{23}$$

The square root of J_2, designated as T_s,

$$T_s = \sqrt{J_2} \tag{24}$$

representing the effective value of shear stresses, accumulated at a given point.

The rock shear strength, Y, limits the value of the computed T_s. The value of Y increases with the local pressure, and decreases with the local temperature. The strength model in the SALEB hydrocode recognises "intact" and "damaged" rocks. "Intact" rocks have a brittle "peak" strength, Y_{int}, at small pressures, gradually transformed to a "ductile" or "plastic" shear limit at pressures of a few GPa. This transition is simulated with a simple analytical approximation, proposed by Lundborg (1968):

$$Y_{int} = Y_o + \frac{k_p \times p}{1 + \frac{k_p \times p}{Y_m - Y_0}} \tag{25}$$

where Y_0 is the shear strength at zero pressure, Y_m is the asymptotic "plastic" shear limit at very high pressure, and k_p is a coefficient to fit experimental data.

In damaged material (dry sand, totally fragmented rocks) the model assumes the dry friction behaviour where the shear strength, Y_{dam}, is linearly proportional to the local pressure:

$$Y_{dam} = f \times p \quad (26)$$

where f is the friction coefficient. The friction coefficient numerical value slightly differs form the usual ratio of the critical shear stress to the normal load, as in the computer model the shear stress state is presented in the form 23–24. The friction value is estimated with fitting of appropriate triaxial test data in the form $T_s = f * p$.

The damage parameter D is used to model he transition from the "intact" state ($D = 0$) to the "damaged" ($D = 1$) state. In the "partially damaged" state, $0 < D < 1$, the strength is assumed to change linearly with D:

$$Y(p) = (1 - D)Y_{int}(p) + DY_{dan}(p) \quad (27)$$

If the "damaged" shear strength reaches the "intact" shear strength, the material is assumed to be in a ductile state when the material at high pressure is deformed like a plastic metal:

$$Y_{dam}(p) = min[f \times p, Y_{int}(p)] \quad (28)$$

The accumulation of the damage D is controlled with the "plastic" strain accumulation (Collins et al. 2004). An example of the SALEB shear strength description is shown in Fig. 2a.

Thermal softening of rocks closing to the melting state is modelled as the shear limit, Y, decreasing with temperature. Ohnaka (1995) proposed a simple analytical expression for $Y(T)$, based of laboratory tests of Westerly granite (Griggs et al. 1960; Stesky et al. 1974; Wong 1982):

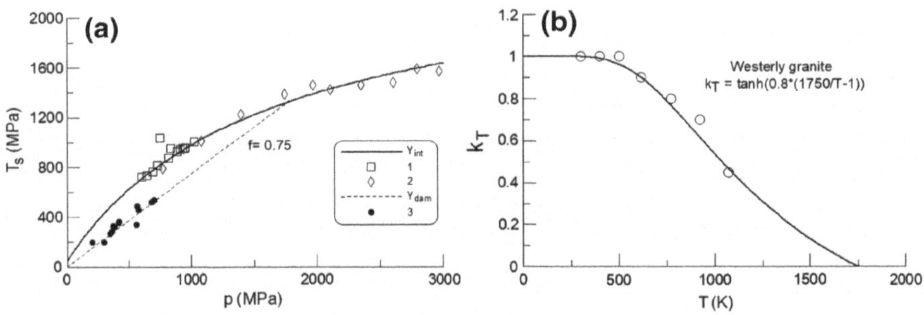

Fig. 2 **a** The strength curves $T_s = Y(p)$ for Westerly granite and similar rocks: the Y_{int} curve presents Eq. 25, the Y_{dam} curve is the dry friction law (26) with model's friction coefficient $f = 0.75$. Experimental data for the "intact" rock strength at the start of fracturing are published (1) by Stesky et al. (1974) and (2) by Schock and Heard (1974). Dry friction data (3) are published by Stesky et al. (1974) the sawcut Westerly granite samples. **b** The model temperature strength coefficient, k_T, for Eq. 29 approximated with Eq. 30 after Ohnaka (1995)

$$Y(p,T) = Y(p,T_0) \times k_T \tag{29}$$

where $Y(p,T_0)$ is the shear strength at "normal" temperature, and k_T is the temperature softening coefficient

$$k_T = \tan h\left[a_T\left(\frac{T_m(p)}{T} - 1\right)\right] \tag{30}$$

where a_T is an experimentally fitted constant, and $T_m(p)$ is the melting temperature at a local pressure. Originally Ohnaka (1995) treated the T_m value as a fitting parameter not included a pressure dependence. For shock wave modelling we assume that T_m is the solidus temperature, so all the procedure of the thermal softening modelling should be treated as a preliminary one. Equation (30) is based on dry rock tests, and for wet granite the dependence would be more complicated (Odedra et al. 2001). Figure 2b illustrates the rock thermal softening in the model.

The melting (solidus) temperature dependence on pressure, $T_m(p)$, is modelled with so called Simon–Glatzel's approximation (Simon and Glatzel 1929), fitted to published data on rock melting points:

$$T_m(p) = T_m(p_0)\left(\frac{p - p_o}{a_{Sim}} + 1\right)^{\frac{1}{c_{Sim}}} \tag{31}$$

where a_{Sim} and c_{Sim} are "Simon's" constants. This approximation is often used for the description of Earth's interiors (Poirier 2000). So far the description of the incongruent multi-mineral rock melting is not a standard feature of hydrocodes, and the liquid/solid transition trough the partial melting state is mostly estimated a posteriori using standard thermodynamic analysis (Ivanov and Melosh 2003).

More details and examples of the rock strength model are published by Collins et al. (2004).

Temporary dry rock friction reduction (acoustic fluidization model). The analysis of crater sizes at the boundary between simple and complex impact craters resulted in the understanding that the unclear process of the temporary strength and dry friction reduction operates on the Earth and other terrestrial planetary bodies (Melosh 1977, 1989; McKinnon 1978). One of proposed models, namely the acoustic fluidization (AF) model has been adjusted to be used in hydrocodes (Melosh and Ivanov 1999; Wünnemann and Ivanov 2003). The AF model has been fitted to observational data and now widely used to model individual terrestrial craters, such as Vredefort, Sudbury, Chicxulub, West Clearwater and many others (Collins et al. 2002; Ivanov 2005; Morgan et al. 2016; Rae et al. 2017).

The model is based on the assumption that the deforming rock massive can be treated as a system of relatively intact blocks separated with fragmented inter-block layers. In the frame of the AF model blocks are able to oscillate relative to one another. During these oscillations the normal force in inter-block layers could decrease, decreasing the dry friction limiting shear stresses. Consequently, during a fraction of the oscillation period, one block could slide relative to its neighbour. Mathematically one may describe this stick-slip motion as a quasi-viscous motion (Melosh and Ivanov 1999). With some simplification the effective kinematic viscosity ν_{lim} of the near-crater rock motion is controlled with the block size h and the oscillation period T_{osc} (Ivanov and Artemieva 2002):

$$\upsilon_{lim} = c_{af} h^2 / T_{osc} \tag{32}$$

where c_{af} is a model coefficient in the range of 4–8. Ivanov and Kostuchenko (1997) analysed available data for the Puchezh-Katunki structure, and concluded that the optimal AF model parameters are h in the range of 100–150 m, T_{osc} in the range of 0.5–1 s, and the effective kinematic viscosity is about 10^5 m^2/s.

The block oscillation's amplitude is assumed to decay in time due to inelastic oscillation energy loss and emanating of the acoustic energy beyond the near-crater region. The exponential decrease of internal oscillations is described in the AF model with the decay time T_{dec}. The ratio of the oscillation period to the decay time presents the quality factor, a measure of the oscillation energy decay per one cycle:

$$Q = T_{osc} / T_{dec} \tag{33}$$

The larger is quality factor, the longer is the whole time of the AF activity. For the Puchezh-Katunki case the suitable value of Q was found to be in the range of 50–100. It means that assumed blocks should oscillate 50–100 times before the internal oscillation decay to a low level, where the dry friction between blocks return to normal values.

In the computer code the AF action is simulated in a few steps:

1. The code record the maximum velocity in each Lagrangian particle during the shock wave passage through the damaged zone around a crater. The accumulation time is comparable with the time to reach the maximum transient cavity depth. When this time, T_{off}, is reached, the maximum velocity is fixed, and the AF model starts to operate.
2. The initial internal oscillation amplitude (in terms of the maximum velocity in the oscillation cycle, v_{vib}) is calculated as a fraction of the recorded maximum particle velocity. This fraction is nominated as an input value c_{vib}. The value of c_{vib} typically varies from 0.1 to 0.3.
3. The additional "pressure" from internal oscillations is estimated in each computational cell as

$$p_{vib} = c_{snd} \rho \, \upsilon_{vib} \tag{34}$$

where c_{snd} is the local sound speed, and ρ is the local rock density.
4. In the stress deviator program section the code first computes the local stress level, assuming the elastic behaviour of rocks, and then check if stresses at the current time step are above or below the local strength level, *Y(p)*. In the normal

case (no AF) the local strength is computed in a standard way via the assumed cohesion Y_0 and the friction coefficient *fric*

$$Y(p) = Y_0 + fric \times p \tag{35}$$

The AF model computes the Bighham-like shear strength, $Y_B(p,\ p_{vib},\ e')$ as a function of local values of pressure, *p*, "vibrational" pressure, p_{vib}, and shear strain rate, e' (Melosh and Ivanov 1999):

$$Y_B(p,\ p_{vib},\ e') = frict \times (p - p_{vib}) + \rho \times e' \times v_{lim} \tag{36}$$

with the friction coefficient *frict*, $\tau \sim frict\,(p - p_{vib})$, and the quasi-viscous shear stress τ_{vis} The code chooses the minimum value of the shear strength without AF, *Y(p)*, and with AF, $Y_B(p,\ p_{vib},\ e')$, and compares the computed value, Y_{loc}, with the value of the local shear, expressed as $Ts = \sqrt{J_2}$ (see Eq. 24). If $Ts \leq Y_{loc}$ (stresses are below the strength limit), the cell material is treated as an elastic one, if $Ts > Y_{loc}$ (stresses are above the strength limit), each stress tensor component is decreasing to stay at the strength surface $Ts = Y_{loc}$. If the reduction of stresses occurs, the deformation in the given cell at the current time step is treated as inelastic one. The inelastic deformation (as the invariant value of strain) is accumulated in a given Lagrangian particle, producing the very informative value—the equivalent inelastic ("plastic") strain *epl_eq*, used to illustrate the final inelastic strain—relative neighbour block displacement. If we assume a block size *h*, inelastic deformations *epl_eq* ~ 1 manifest that one block accumulated a displacement ~*h* relative to the neighbour block. In these terms the value of *epl_eq* serves as a measure of a structure disordering. Zones with *epl_eq* < <1 seem to preserve relative structural features (like layering etc.), while zones with *epl_eq* ≥ 1 may be treated as candidates for mixing of the target material blocks.

Values, v_{vib} and *epl_eq* are advected with other "Lagrangian" parameters through the Eulerian computational grid.

2 Computation and Geological Data

Before to present the Puchezh-Katunki numerical model, we should make a "disclaimer": the numerical model we use is able to reproduce only a limited set of physical and mechanical processes during the impact crater formation. In the presented work we do not model the atmospheric entry and the influence of the atmosphere at ejecta motion—see examples of more advanced modelling by Artemieva et al. (2013); Shuvalov and Dypvik (2013), and many others. We do not model a possible shallow water layer at the Puchezh-Katunki site at the time of formation, and the water saturation of the upper sedimentary layers. Examples of the numerical modelling with a water layer and water (ice) saturated targets have been published by Lindström et al. (2005), Ivanov and Pierazzo (2011), and many others. We do not model the most probable case of an oblique impact—3D modelling of simple and complex crater formations has been started in the last 10 years (Elbeshausen et al. 2009, 2013; Wünnemann and Elbeshausen 2010), but still demands much larger spatial resolution The interested reader is invited to read about new modelled mechanisms elsewhere.

We have concentrated in the numerical modelling of a simplified impact scenario of the Puchezh-Katunki structure formation, namely—the modelling of a vertical impact into a target where dry sedimentary layer of 2 km thick overly the crystalline basement. The main goals where (1) to reproduce the final crater morphologically similar to the PK structure, and (2) to verify the modelling with estimated maximum shock pressure in uplifted rocks by comparison with samples, recovered from the VDW.

The initial guess for suitable projectile mass and impact velocity has been done in early attempts to simulate the Puchezh-Katunki crater formation (Ivanov 2002, 2008). The impact velocity range for asteroid impacts on Earth is from 11.2 to ~30 km s^{-1} with the average value of 18–20 km s^{-1} (Werner and Ivanov 2015). In the presented study an important issue is the shock pressure, recorded in rocks and minerals. For this reason, we should take into account that the vertical impact gives the maximal strong shock wave, while oblique impacts tends to generate weaker shock wave. At the initial stage the shock wave in a target depends mostly on the vertical impact velocity component (Pierazzo and Melosh 1999). To mimic an oblique impact with the available vertical impact model, we make most of model runs with impact velocities from 10 to 16 km s. Larger impact velocities for the vertical impact result in the voluminous massive impact melt not observed in the Puchezh-Katunki structure. Also, we made an attempt to reproduce an elongated source zone for an oblique impact changing the usual spherical projectile with an elliptical body.

The limited set of available equations of state include Tillotson's EOS for sediments ("wet tuff") and crystalline basement ("granite") (Melosh 1989), and the ANEOS-based computational tables for granite (Pierazzo et al. 1997).

Figure 3 illustrate the sequence of main consequent phases, characterizing the high velocity impact cratering. The starting geometry is shown in Fig. 3a where vertically travelled (downward) spherical asteroid touches the surface of a target. The impact generates a shock wave in the target, the transient cavity begins to open, and near-surface material is ejected up—and outward. Approximately in 10 s (the lower panel in Fig. 3a) the transient cavity reaches its maximum depth, continuing to expand in diameter.

Near the time of the maximum transient cavity depth (the lower panel in Fig. 3a) the code "switches on" the AF model, and the dry friction under the cavity decreases. Consequently, the transient cavity floor starts to rise up (resembling the water cavity collapse after a pebble falling in a pond). Approximately at 30 s (the upper plate in Fig. 3b), the ejection of the target material ends, and the vertical velocity component at the crater edge at the preimpact level changes direction -from upward ("ejection") to downward ("collapse"). The cavity radius at this time moment may be conditionally named as "the transient cavity radius" despite the cavity radius continues to grow, but due to the collapse continuation.

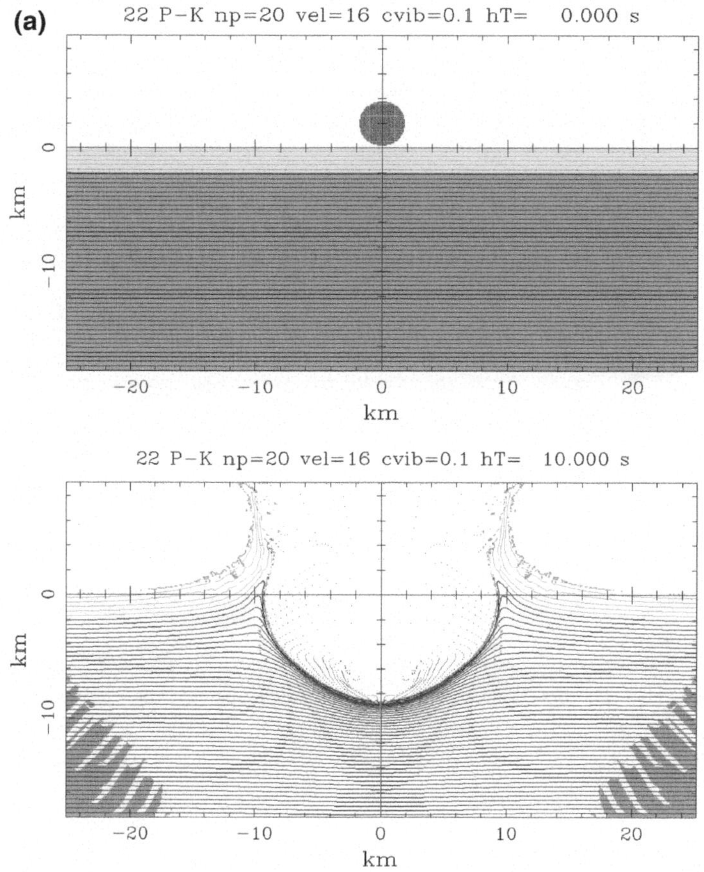

Fig. 3 **a** Two frames (initial and 10 s) frames of the crater formation sequence. **b** Two more frames (at 30 and 80 s) the crater formation sequence. The dark grey colour near the center mark the basement material above the melting temperature. The gray color at the periphery shows unfractured basement rocks, designating the rock rupture zone boundary. **c** Two final frames (at 84 and 156 s) the crater formation sequence. About 80 s after impact the upward collapse flow ends, and after a small downward collapse the crater reach the final shape. Horizontal curves show deformations of initially horizontal layers of Lagrangian tracers. Darker curves are for crystalline basement, lighter curves are for sedimentary rocks, occupied initially the upper 2 km of the target

Due to the temporary low dry friction, described with the AF model, the central sub-crater rock massive starts to move in a "liquid" style—marginal part subsides in the gravity field, pushing up the central part. This motion creates the *central uplift*: rocks near the axis of axial symmetry, pushed to the depth of 8–9 km below the surface, move up to the preimpact level. In a real liquid target the central uplift would create an upward jet, but due to the AF effect decay in time, the dry friction slowly returns to the normal value and decelerates the uplift growth. We should note, that the appropriate decay time (40–45 s) is comparable with the time period between

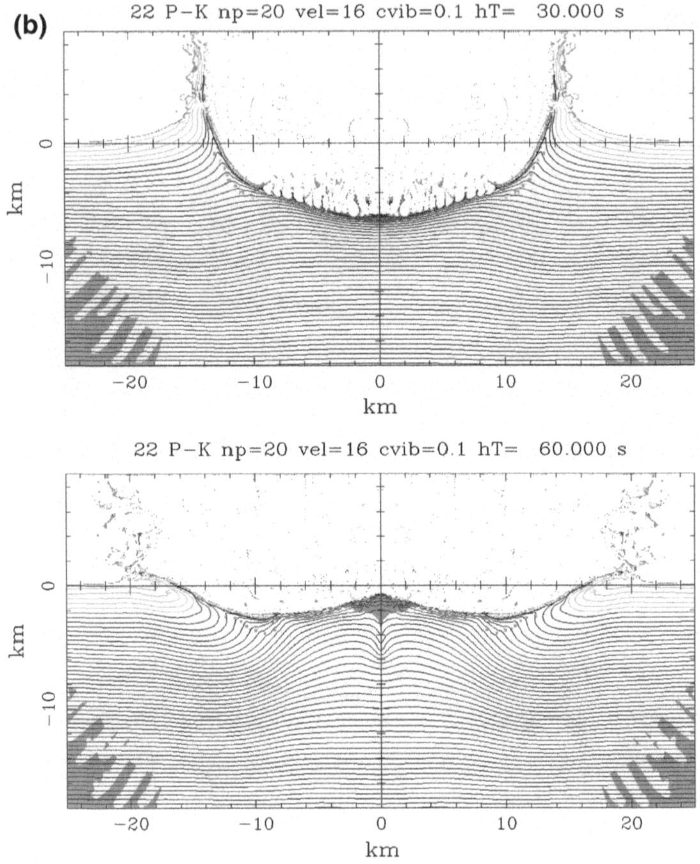

Fig. 3 (continued)

the upward motion start (t ~ 10 s after impact, the lower panel in Fig. 3a) and stop (t ~ 80 s after impact, the upper panel in Fig. 3c).

The main feature of the numerical model is the prediction of the initial depth of rocks, uplifted close to the surface in the central mound. For the Puchezh-Katunki crater size, the net uplift is 4–6 km. The exact shape of distorted horizontal layers depends on some interaction of the input parameter's set (and will be illustrated later in the text).

The minor collapse of the slightly overshot central uplift finalises the complex impact crater formation. This general scheme could slightly change in details while the modeler varies model parameters, and we still have no recipes to nominate values of multiple parameters in advance. The parameter fit is the essential part of the modelling—see an example of the proper parameter search by Rae et al. (2017) for the West Clearwater Lake impact crater formation.

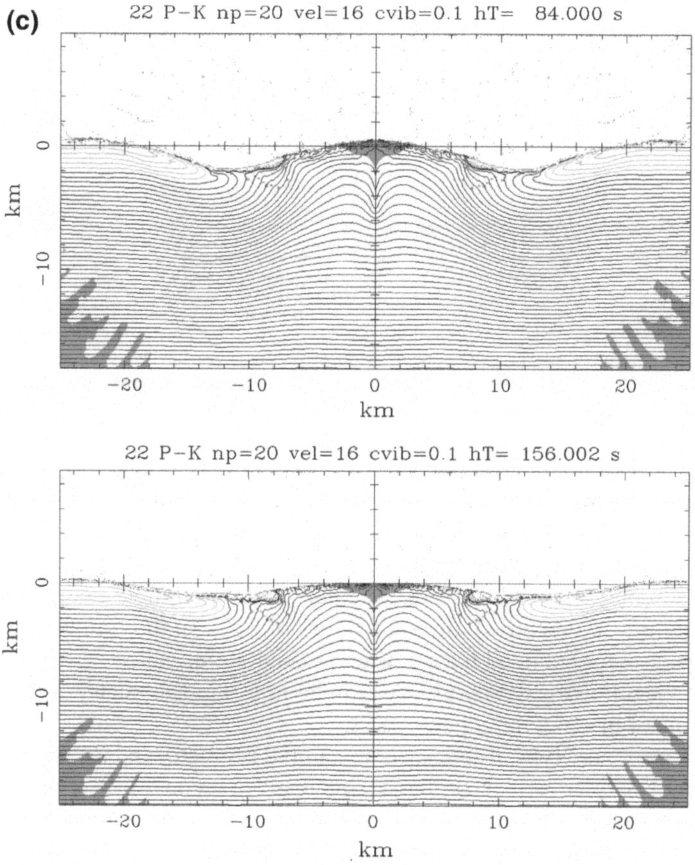

Fig. 3 (continued)

The most robust part of the numerical model is the shock wave propagation and decay. The shock pressure decay along the axis of symmetry for a set of model runs is shown in Fig. 4.

All models with the projectile size, fitted to produce a final crater with D $\approx$ 40 km, produce shock waves of 100–200 GPa at the basement upper boundary and decay to ~10 GPa at the depth of ~10 km.

To present the shock wave pressure distribution in space out of the symmetry axis, two kind of representations are commonly used. Both are based on the usage of massless tracers, initially located at the centre of each computational cell. These tracers move through the Eulerian computational grid as Lagrangian particles, and play role of "pebbles", recorded the pressure-temperature history during all stages of a crater formation. With some limited accuracy, we can treat these tracers as real rock fragments, shocked and displaced in the area of the dill core sampling. Looking back in time, we can (approximately) restore where the given sample was located initially

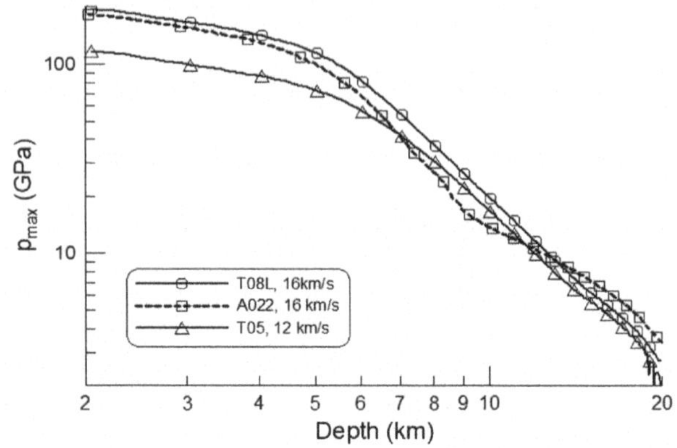

Fig. 4 The pressure decay in the crystalline basement for three model runs with different impact velocities and equations of state

and what was the maximum shock strength in this location. The first representation form is to plot the maximum shock pressure value, p_{max} (more exactly—the pressure jump above the initial lithostatic pressure), using initial coordinates of our tracers. The second form is the direct plotting of p_{max} in final coordinates of tracers. In the Eulerian computational grid, in the final position a single computational cell may contain tracers, recorded different values of p_{max}. In the following figures we averaged p_{max} recorded in all tracers, accumulated in this cell.

A set of the following figures gives examples of general geometry of shock isobars for three sets of model's input parameters.

Figure 5 presents a typical shock isobar's initial geometry and its distortion during the transient cavity growth and collapse. In the given model run the shape of 10GPa isobar is only slightly distorted, and its maximum depth at the symmetry axis is about to be preserved. 20 and 30 GPa isobars are heavily distorted, and at the centreline x = 0 rocks, shocked to these pressures, are uplifted ~4–6 km above the initial depth. 40 GPa isobar predicts an average recorded pressure to the central part of the crater mound. Rocks out of 8–10 km from the center are shocked well below 10 GPa level.

Examples of impacts with higher velocity of 16 km s^{-1} are shown in Fig. 6 for Tillotson's EOS and in Fig. 7 for the ANEOS "granite" (Pierazzo et al. 1997).

The important point should be taken into account: the initial geometry of shock pressure isobars (left panels in Figs. 5, 6 and 7) depends on physics of the shock wave propagation, controlled with the material equation of state and with wave refraction at the free surface. The final isobar geometry (right panels in Figs. 5, 6 and 7) is controlled with the material displacement during the transient cavity collapse. In the current logic of the AF model (defines the local friction coefficient) the initial internal oscillation amplitude is proportional to the maximal particle velocity at the shock front. For this reason, the material EOS could affect the initial strength of the AF effect. This relatively small AF variability would change the displacement geometry, resulting in visible difference of shock isobar's geometry after the complex crater formation.

We also check how the assumed shape of the projectile could affect the crater formation and shock isobar's geometry. Figure 8 compares two model runs where

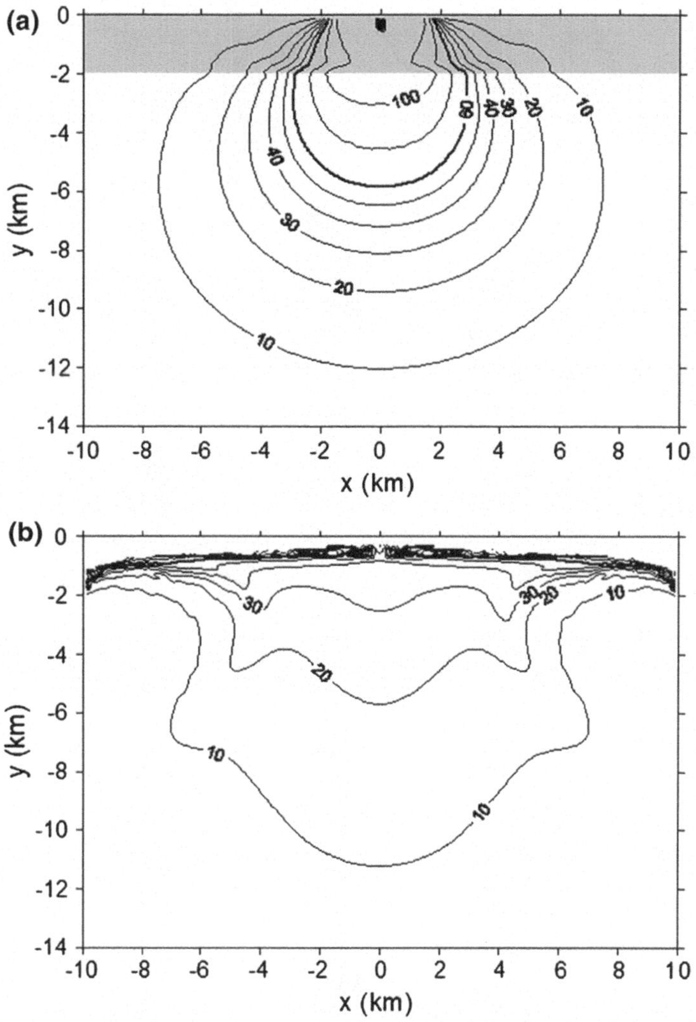

Fig. 5 Initial **a** and final **b** geometry of shock pressure isobars (labelled in GPa) for Tillotson's EOS and impact velocity of 12 km s^{-1}. The thick curve 60 GPa designates the approximate boundary of the impact melt zone. After the crater collapse the melted material is pushed to the surface and mainly not resolved with the 100 × 100 m computational cells

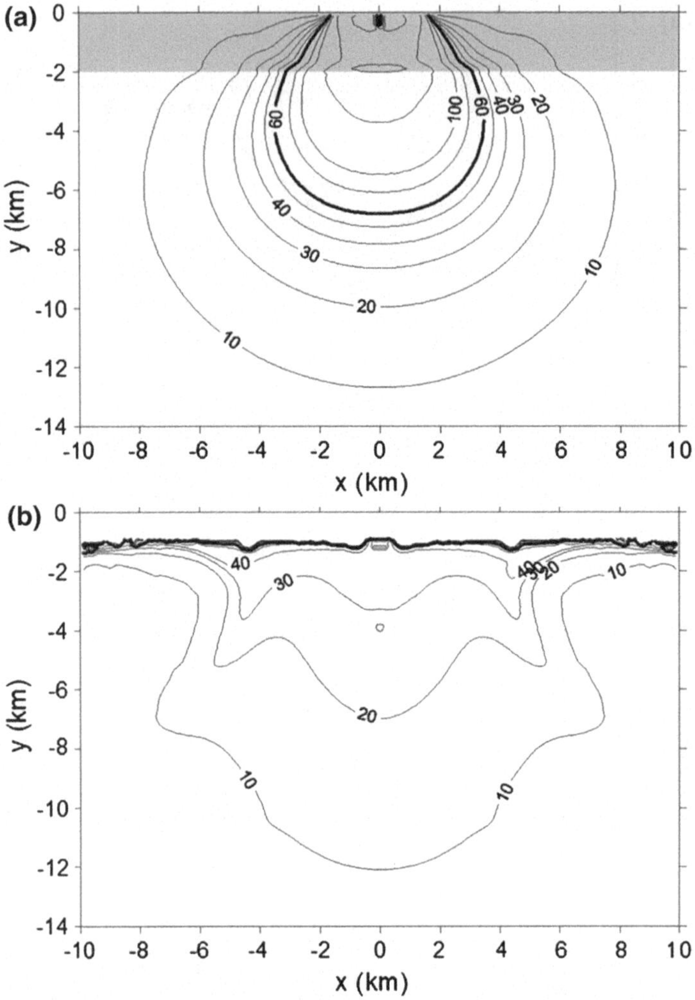

Fig. 6 Initial **a** and final **b** geometry of shock pressure isobars (labelled in GPa) for Tillotson's EOS and impact velocity of 16 km s^{-1}. The thick curve 60 GPa designates the approximate boundary of the impact melt zone. After the crater collapse the melted material is pushed to the surface and mainly not resolved with the 100 × 100 m computational cells

equal mass projectiles, with a spherical shape and an elliptical shaped, strike the target surface. With a minor difference material displacement's geometry is the same for both projectiles.

Shock pressure isobar's geometries for two model runs (Fig. 8) are shown in Fig. 9 in a way similar to Figs. 5, 6 and 7. Interestingly the 60 GPa shock pressure isobar (characteristic impact melting pressure) reaches the depth of ~6 km for the elliptical projectile impact (Fig. 9c), while for spherical projectile the maximum depth is only ~4 km (Fig. 9a). The difference seems to result from a stronger pressure decay in

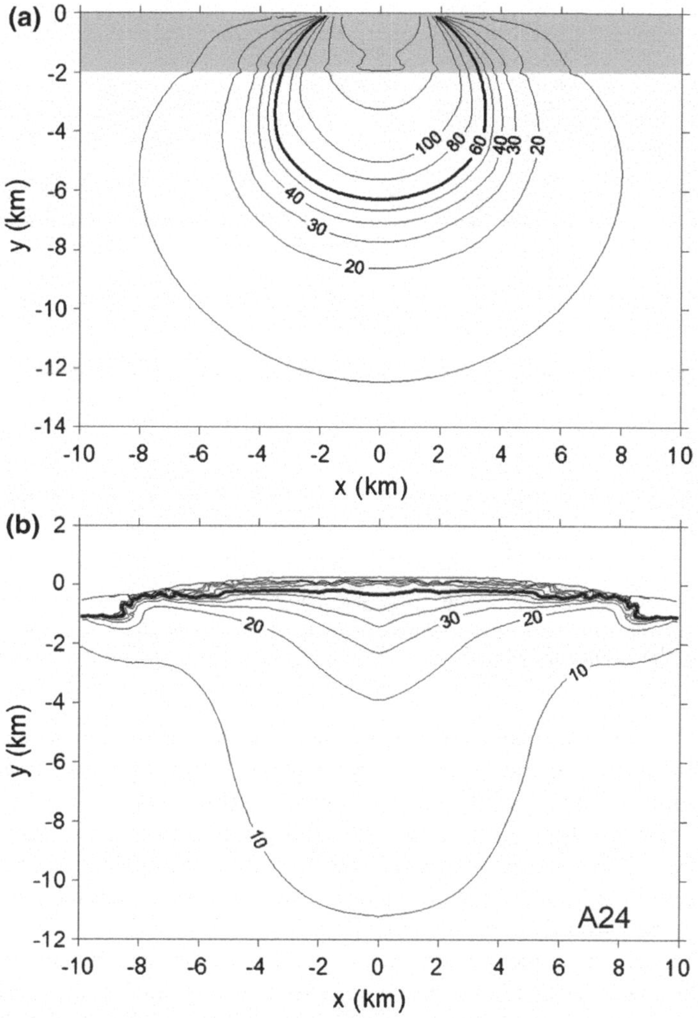

Fig. 7 The same as in Fig. 8.4, but for the model run with the ANEOS "granite" model for the basement rocks

sedimentary rocks. The difference in the maximum shock pressure (and maximum particle velocities, influencing the AF oscillation strength) is finally reflected in the final 20 GPa isobar position (Fig. 9b and c).

We compare below model predictions and actually observed shock load level, recorded in recovered VDW core samples (see Chap. 4). Figure 10 presents shock pressures recorded in tracers displaced to the model "drill hole"—a vertical tube near the crater center—for two model runs with the impact velocity of 16 km s^{-1} (the right panel) and with the low impact velocity of 10 km s^{-1} (the left panel). The vertical axis in this figure is the depth below the pre-impact surface, the horizontal

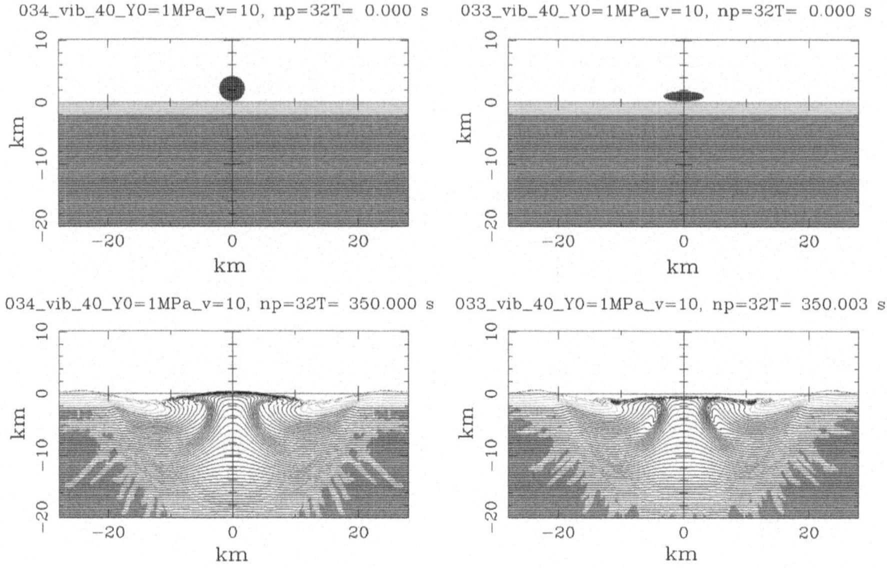

Fig. 8 Impact of a spherical projectile (left panels) and of an elliptical projectile (right panels). Both projectiles have the same mass and the same impact velocity of 10 km s^{-1}. Grey shading in the basement shows undamaged rocks

axis shows the recorded shock pressure for tracers at the given depth. Model data are compared with mineralogical estimates of the maximum shock pressure recorded in samples, collected at various depths from the VDW (Fig. 4.1).

The comparison shows the relatively good coincidence of observed and modelled shock pressure estimates. However, the whole set of tested model variants reveals a kind of instability of final results. It seems to result from oversimplification of many elements in the currently available models.

While we model simple impact craters, where exotic models like the AF are not needed, the main verification is the comparison of modelled and experimentally measured crater profiles—the instructive example in (Wünnemann et al. 2016). For complex crater formation the usual problem of a limited spatial resolution of the model/experiment comparison makes the fit not so convincing as in the case of simple craters—see, for example, the instructive example of Chicxulub numerical modelling (Morgan et al. 2016). In the presented Puchezh-Katunki numerical modelling the spatial resolution is 100 m (quadrangle computational cell 100 × 100 m), and the model should cover the shock wave propagation 50–100 km out of the point of impact, the transient cavity with a maximum depth of 10 km, and the final crater shape with the depth of the order of 1 km. Finally, the crater profile, drawn through the boundary cells between vacuum (or vaporized material) and the condensed material (solid or liquid) is relatively rough. Another complexity is the poor knowledge of the real fresh impact crater profile before millions of years of erosion and denudation. Having in mind all these problems we use the geologic reconstruction along SW-VDW profile,

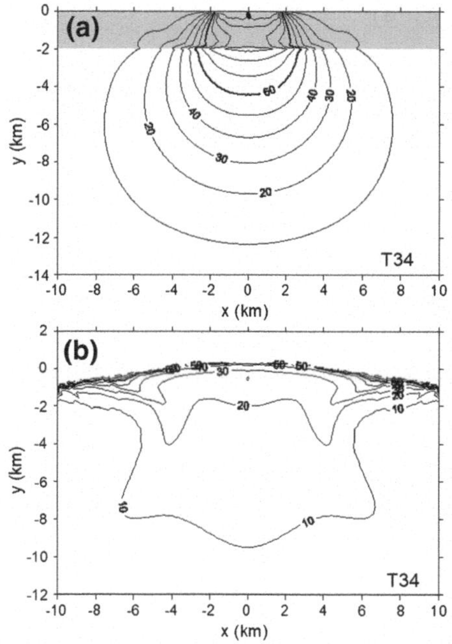

Fig. 9 Geometry of shock pressure's isobars for the spherical projectile (a—initial position, b—final position), and for the elliptical projectile (c—initial position, d—final position) of equal masses. Tillotson's EOS, impact velocity 10 km s^{-1}

where numerous drill holes allowed us to restore the boundary between crater-related rocks and younger sediments (Fig. 3.3). Figure 11 compares boundaries, digitized from Fig. 3.3, with two model runs, consistent with core-based p_{max} (z) data.

The model was initially targeted to fit the crater diameter of ~40 km (20 km radius) at the preimpact surface. The depth in both model variants is about 0.8–1 km, what is compatible with observations taking into account 100 m computational cell size. Having a general similarity with the real crater surface topography, the model does not take into account the possible rim outflow motion, recorded in the radial grooves (see Sect. 3.1). The possible Martian analogues of the post-deposition outward rim material motion has been modelled in (Ivanov 1996).

The direct numerical modelling allows us to outline the range of size and kinetic energies of an asteroid able to create the Puchezh-Katunki impact structure. Within impact velocity range of 11–16 km s^{-1} the probable asteroid mass varies from ~5 × 10^{13} to ~10^{14} kg with the kinetic energy from ~5 × 10^{21} to ~10^{22} J. At the "granite" density of 2600 kg m^{-3}, the equivalent spherical diameter varies from ~3.5–4 km. These estimates could vary for oblique impacts. Earth's atmosphere is unable to decelerate such massive bodies. However, the presence of an atmosphere (not modelled here) would affect ejecta motion, resulted in dust cloud deposition inside the just formed crater.

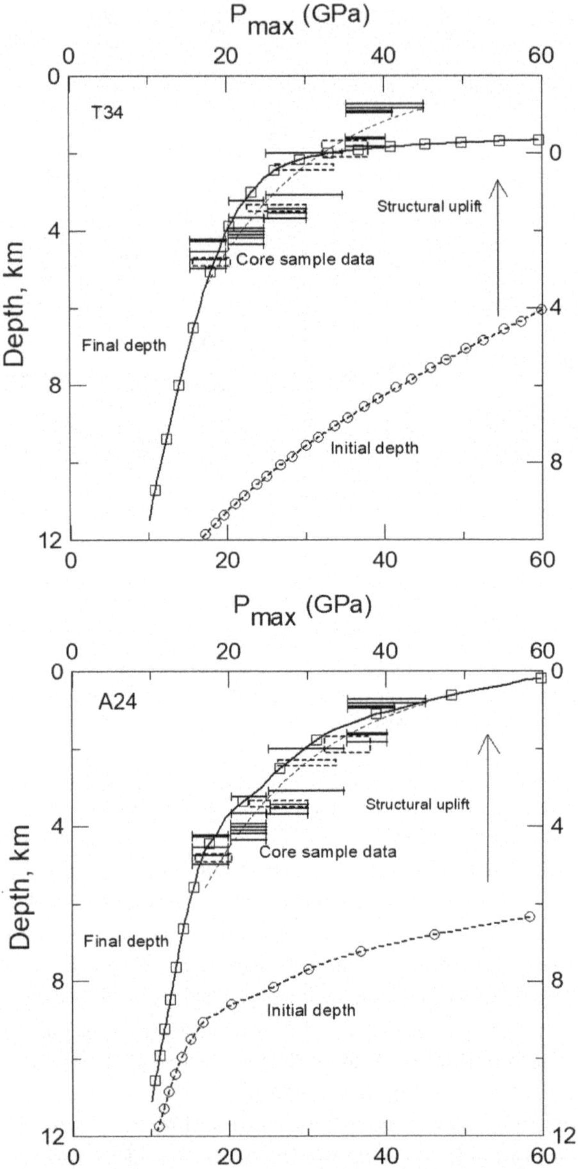

Fig. 10 The comparison of recorded maximal shock pressure, p_{max}, values versus sample depth, z, from rock mineral shock transformations (labelled "Core sample data") and two model variants. Model data are presented as p_{max} (z) where z is the initial depth below the pre-impact surface (curves with circle signs) and as the p_{max} (z) where z is the final depth after the transient crater collapse. As the exact position of the real pre-impact surface is unknow, model curves are fitted to measured points by slight vertical shift up and down. The shifted model depths are shown at the right vertical axis. Left panel—the low velocity (10 km s^{-1}) impact of an elliptical projectile with the equivalent sphere diameter of 4 km, Tillotson's EOS. Right panel—higher velocity impact (16 km s^{-1}) impact of a spherical projectile, 3.6 km in diameter, ANEOS "granite"

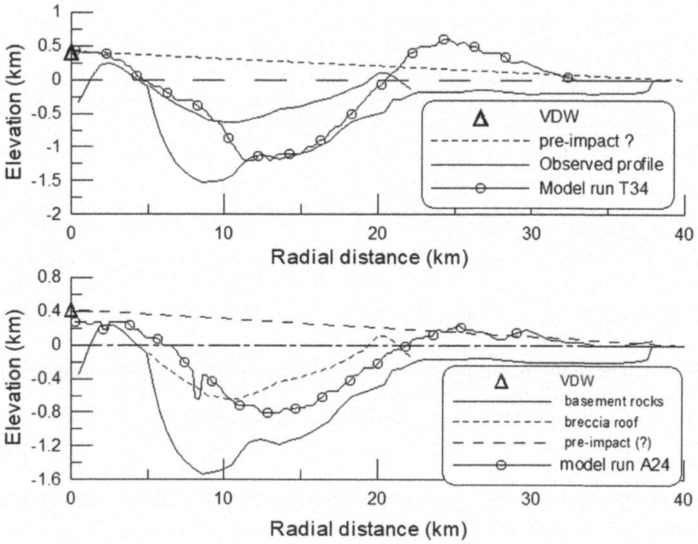

Fig. 11 Modelled final crater profiles (curves with circle signs) in comparison with geologically estimates along SW-VDW profile (Fig. 3.3). The upper panel is for a slow impact, the lower panel is for larger impact velocity

Concluding the numerical model presentation, one should note that the model ability to reproduce both the deep transient cavity and the shallow final complex crater profile is a prominent achievement of the AF model, implemented about 20 years ago in hydrocodes. However, the experience of the AF model usage allows us to see problems in the perception of the model after accumulation of new observational data, including deep drilling in terrestrial craters. Various authors propose to implement the variable block size in space and time, a possibility to generate additional internal oscillations due to shear deformation of blocks and others (Collins and Melosh 2002; Riller et al. 2018). In a future, the ground truth of the Puchezh-Katunki investigations and the material recovered in VDW will be definitely use to construct better numerical models and to re-think data about thousands of similar impact structures on terrestrial planets, satellites, and asteroids.

References

Allen RT (1967) Equation of state of rocks and minerals. Interrim report to DASA under contract #DA49-146-XZ-462 l, General Dynamics Report GAMD-7834, San Diego, California, 23 pp

Amsden AA, Ruppel HM, Hirt CW (1980) Sale: a simplified ALE computer program for fluid flow at all speeds. Los Alamos Laboratory Report LA-8095 l, Los Alamos, NM, 101 pp

Anderson DL (2005) New theory of the earth. Cambridge University Press, Cambridge, UK, p 408

Artemieva NA, Wünnemann K, Krien F, Reimold WU, Stöffler D (2013) Ries crater and suevite revisited—observations and modeling part II: modeling. Meteorit Planet Sci 48(4):590–627

Bosslough MB (1988) Postshock temperatures in silica. J Geophys Res 93:6477–6484
Brundage AL (2013) Implementation of Tillotson Equation of State for Hypervelocity Impact of Metals, Geologic Materials, and Liquids. Procedia Eng 58:461–470
Collins GS, Melosh HJ (2002) Target weakening and temporary fluidization in large impact events (abs.). XXX Lunar and Planetary Science. Houston, TX, # 1523
Collins GS, Melosh HJ, Morgan JV, Warner MR (2002) Hydrocode simulations of chicxulub crater collapse and peak-ring formation. Icarus 157:24–33
Collins GS, Melosh HJ, Ivanov BA (2004) Modeling damage and deformation in impact simulations. Meteorit Planet Sci 39(2):217–231
Do IPH, Benson DJ (2000) Modeling shock-induced chemical reactions. Int J Comput Eng Sci 1(01):61–79
Elbeshausen D, Wünnemann K, Collins GS (2009) Scaling of oblique impacts in frictional targets: implications for crater size and formation mechanisms. Icarus 204:716–731
Elbeshausen D, Wünnemann K, Collins GS (2013) The transition from circular to elliptical impact craters. J Geophys Res (Planets) 118:2295–2309
Griggs DT, Tumer FJ, Heard HC (1960) Deformation of rocks at 500–800 °C. In: Griggs D, Handin J (eds) Rock deformation. Geol. Soc. Amer. Memoir 79, Waverly Press, Baltimore, MD, pp 39–104
Ivanov BA (1996) Spread of ejecta from impact craters and the possibility of estimating the volatile content of the martian crust. Sol Syst Res 30:36–52
Ivanov BA (2002) Deep drilling results and numerical modeling: puchezh-katunki impact crater, Russia (abs.), XXX lunar and planetary science. Houston, TX, # 1286
Ivanov BA (2005) Numerical modeling of the largest terrestrial meteorite craters. Sol Syst Res 39:381–409
Ivanov B (2008) Geologic effects of large terrestrial impact crater formation. In: Adushkin VV, Nemchinov IV (eds) Catastrophic events caused by cosmic objects. Springer, Berlin, pp 163–205
Ivanov BA, Artemieva NA (2002) Numerical modeling of the formation of large impact craters. In: Koeberl C, MacLeod KG (eds) Catastrophic events and mass extinctions: impact and beyond. Geological Society of America Spec. Pap. 356, GSA, Boulder, Colorado, pp 619–630
Ivanov BA, Kostuchenko VN (1997) Block oscillation model for impact crater collapsel (abs.). XXVIII Lunar and Planetary Science. Houston, TX, pp 631–632
Ivanov BA, Melosh HJ (2003) Impacts do not initiate volcanic eruptions: Eruptions close to the crater. Geology 31(10):869–872
Ivanov BA, Pierazzo E (2011) Impact cratering in H_2O-bearing targets on mars: thermal field under craters as starting conditions for hydrothermal activity. Meteorit Planet Sci 46:601–619
Ivanov BA, Basilevsky AT, Neukum G (1997a) Atmospheric entry of large meteoroids: implication to Titan. Planet Space Sci 45:993–1007
Ivanov BA, Deniem D, Neukum G (1997b) Implementation of dynamic strength models into 2D hydrocodes: applications for atmospheric breakup and impact cratering. International Journal of Impact Engineering, Hypervelocity Impact Proceedings of the 1996 Symposium, vol 20, no1–5, pp 411–430
Ivanov BA, Langenhorst F, Deutsch A, Hornemann U (2002) How strong was impact-induced CO2 degassing in the Cretaceous-Tertiary event?: numerical modeling of laboratory experiments. In: Koeberl C, MacLeod KG (eds.) Catastrophic events and mass extinctions: impact and beyond. Geological Society of America Spec. Pap. 356, GSA, Boulder, Colorado, pp 587–594
Ivanov BA, Melosh HJ, Pierazzo E (2010) Basin-forming impacts: Reconnaissance modeling. In: Gibson RL, Reimold WU (eds) Geol Soc Amer Special Papers 465, Geological Society of America, Boulder, Colorado, USA, 2010, pp 29–49
Kurosawa K, Genda H (2018) Effects of friction and plastic deformation in shock-comminuted damaged rocks on impact heating. Geophys Res Lett 45:620–626
Lindström M, Shuvalov V, Ivanov B (2005) Lockne crater as a result of marine-target oblique impact. Planet Space Sci 53(8):803–815

Lundborg N (1968) Strength of rock-like materials. Int J Rock Mech Min Sci and Geomech 5(5):427–454
McGlaun JM, Thompson SL, Elrick MG (1990) CTH: a three-dimensional shock wave physics code. Int J Impact Eng 10(1-4):351–360
McKinnon WB (1978) An investigation into the role of plastic failure in crater modification. In: Proceeding lunar and planetary science conference 9th (Geochimica et Cosmochimica Acta. Supplement 10), pp 3965–3973
Melosh HJ (1977) Crater modification by gravity—a mechanical analysis of slumping. In: Roddy DJ, Pepin RO, Merrill RB (eds) Impact and explosion cratering: planetary and terrestrial implications. Pergamon Press, NY, pp 1245–1260
Melosh HJ (1989) Impact cratering—a geologic process. Oxford Univ. Press, Oxford - New York, p 245
Melosh HJ (2007) A hydrocode equation of state for SiO2. Meteorit Planet Sci 42:2079–2098
Melosh HJ, Ivanov BA (1999) Impact crater collapse. Annu Rev Earth Planet Sci 27:385–415
Melosh HJ, Ivanov BA (2018) Slow impacts on strong targets bring on the heat. Geophys Res Lett 45:2597–2599
Melosh HJ, Ryan EV, Asphaug E (1992) Dynamic fragmentation in impacts: Hydrocode simulation of laboratory impacts. J Geophys Res 97:14735–14759
Morgan JV, Gulick SPS, Bralower T, Chenot E, Christeson G, Claeys P, Cockell C, Collins GS, Coolen MJL, Ferrière L, Gebhardt C, Goto K, Jones H, Kring DA, Le Ber E, Lofi J, Long X, Lowery C, Mellett C, Ocampo-Torres R, Osinski GR, Perez-Cruz L, Pickersgill A, Poelchau M, Rae A, Rasmussen C, Rebolledo-Vieyra M, Riller U, Sato H, Schmitt DR, Smit J, Tikoo S, Tomioka N, Urrutia-Fucugauchi J, Whalen M, Wittmann A, Yamaguchi KE, Zylberman W (2016) The formation of peak rings in large impact craters. Science 354(6314):878–882
Murnaghan FD (1944) The compressibility of media under extreme pressures. Proc Natl Acad Sci 30:244–247
Odedra A, Ohnaka M, Mochizuki H, Sammonds P (2001) Temperature and pore pressure effects on the shear strength of granite in the brittle-plastic transition regime. Geophys Res Lett 28:3011–3014
Ohnaka M (1995) A shear failure strength law of rock in the brittle-plastic transition regime. Geophys Res Lett 22:25–28
Pierazzo E, Melosh HJ (1999) Hydrocode modeling of Chicxulub as an oblique impact event. Earth Planet Sci Lett 165:163–176
Pierazzo E, Vickery AM, Melosh HJ (1997) A reevaluation ofimpact melt production. Icarus 127:408–423
Poirier J-P (2000) Introduction to the physics of the Earth's interior. Cambridge University Press, Cambridge, UK, p 326
Rae ASP, Collins GS, Grieve RAF, Osinski GR, Morgan JV (2017) Complex crater formation: Insights from combining observations of shock pressure distribution with numerical models at the West Clearwater Lake impact structure. Meteorit Planet Sci 52:1330–1350
Raikes SA, Ahrens TJ (1979) Post-shock temperatures in minerals. Geophys J Roy Astron Soc 58(3):717–747
Riller U, Poelchau MH, Rae ASP, Schulte FM, Collins GS, Melosh HJ, Grieve RAF, Morgan JV, Gulick SPS, Lofi J, Diaw A, McCall N, Kring DA, Party I-IES (2018) Rock fluidization during peak-ring formation of large impact structures. Nature 562:511–518
Schock RN, Heard HC (1974) Static mechanical properties and shock loading response of granite. J Geophys Res 79:1662–1666
Shuvalov VV (1999) Multi-dimensional hydrodynamic code SOVA for interfacial flows: Application to the thermal layer effect. Shock Waves 9(6):381–390
Shuvalov V, Dypvik H (2013) Distribution of ejecta from small impact craters. Meteorit Planet Sci 48:1034–1042
Simon F, Glatzel G (1929) Bemerkungen zur Schmelzdruckkurve. Zeitschrift fьr anorganische und allgemeine Chemie 178(1):309–316

Stesky RM, Brace WF, Riley DK, Robin P-YF (1974) Friction in faulted rock at high temperature and pressure. Tectonophysics 23:177–203

Thompson SL, Lauson HS (1972) Improvements in the Chart-D radiation hydrodynamic code III: revised analytical equation of state. Sandia Laboratories, Albuquerque, NM, SC-RR-71 0714, 0119 pp

Tillotson JH (1962) Metallic equations of state for hypervelocity impact. In: General Atomic Report GA-3216, San Diego, CA, 141 pp

Trunin RF, Gudarenko LF, Zhernokletov MV, Simakov GV (2001) Experimental data on shock compression and adiabatic expansion of condensed matter. RFNC-VNIIEF, Sarov, Russia, p 446

Watts AW, Greeley R, Melosh HJ (1991) The formation of terrains antipodal to major impacts. Icarus 93:159–168

Werner SC, Ivanov BA (2015) Exogenic dynamics, cratering, and surface ages (Chapter 10.10). In: Schubert G (ed) Treatise on geophysics, 2nd edn. UK, Oxford, pp 327–365

Wong TF (1982) Shear fracture energy of Westerly granite from post-failure behavior. J Geophys Res 87:990–1000

Wünnemann K, Elbeshausen D (2010) Structural evidence for oblique meteorite impacts at complex crater structures from 3D numerical modelling (abs.). Proc. European Planetary Science Congress 2010, # 74

Wünnemann K, Ivanov BA (2003) Numerical modelling of the impact crater depth–diameter dependence in an acoustically fluidized target. Planet Space Sci 51(13):831–845

Wünnemann K, Zhu MH, Stöffler D (2016) Impacts into quartz sand: Crater formation, shock metamorphism, and ejecta distribution in laboratory experiments and numerical models. Meteorit Planet Sci 51:1762–1794

Zeldovich IB, Raizer IP (1966–67) Physics of shock waves and high-temperature hydrodynamic phenomena. Academic Press, New York, 488 pp

Zharkov VN, Kalinin VA (1968) Equations of state for solids at high pressure and temperature. Nauka Press, Moscow, p 311 (in Russian)

Zharkov VN, Kalinin VA (1971) Equations of state for solids at high pressures and temperatures. Consultants Bureau, New York, p 257

Conclusion

The Puchezh-Katunki impact structure situated in the center of East European Platform is a unique geological object due to both its origin and its perfect preservation against erosion. The same is true as to the geological knowledge of the structure, first of all due to the extensive drilling (about 125 deep boreholes) including the Vorotilovo Deep Well, which is the world's first deep well (5374 m) drilled in the center of a giant terrestrial impact structure. Besides, it is the only borehole in the Russian Plate that penetrated into the Early Precambrian crystalline basement for a depth of more than 4.8 km. Taking into consideration the formation mechanism of the central uplift in impact craters (its visible elevation is ca. 2–2.5 km, depth of excavation in basement rocks is estimated to be 7–8 km), it may be thought that the crystalline rocks opened by VDW occurred before the impact at a depth of 7–12 km from the present-day surface. Thus, the VDW reached in fact the same depth that the Kola super-deep well.

Detailed geological, geophysical, petrographic, mineralogical, geochemical, and isotope studies made it possible to recognize with confidence interior structure of the crater, composition and distribution of impact-transformed and impact-generated formations.

Three main morphostructural elements are distinguished: (1) outer ring terrace of 15–30 km wide complicated by narrow and shallow radial trough and overlain by a thin blanket variegated allogenic breccia; (2) annular trough with an outer diameter of 40–42 km and depth ca. 1.9–1.6 km, filled with various allochthonous impact-derived rock units (mainly allogenic breccias after sedimentary rocks) overlain by crater lake deposits and more recent sediments; (3) central uplift of 8–10 km in diameter and up to 2 km high with a central pit up to 0.5 km deep. The uplift is composed of blocks of both crystalline basement and sedimentary rocks; it stands out sharply by strong gravity and magnetic highs.

The main lithological units composing the impact structure are: (1) ruptured and deformed crystalline basement and sedimentary cover rocks in the near-crater zone (authigenic breccia); (2) crater fill comprising various allogenic crystalline and sedimentary breccias and impactites (suevites and tagamites); (3) crater lake

V. L Masaitis and M. V Naumov (eds.), *The Puchezh-Katunki Impact Crater*, Impact Studies, https://doi.org/10.1007/978-3-030-32043-0

deposits (including rewashed clastic material of breccias and impactites); (4) overlying deposits (sedimentary rocks distributed throughout the region). According to recent and revised previous isotope and palynological data, the Puchezh-Katunki crater was formed in the Early Jurassic (Sinemurian or Early Pliensbachian, ca. 192–190 Ma).

Impact-induced transformations of crystalline rocks of the central uplift and their lateral and vertical gradients are studied in detail due to extensive drilling, particularly in VDW cores down to about 5 km deep. The crystalline rock underwent shock metamorphism, crushing and displacement, injections of impact melt, local post-shock heating and superposed hydrothermal alteration. Based on shock effects in minerals (including appearance of diaplectic minerals and glasses, coesite and impact diamond), the attenuation of shock pressure is established from 40 to 45 GPa at near-surface zone of the central uplift to 15–20 GPa at the depth of 5 km.

Within some linear zones of intense cataclasis, shocked rocks underwent a thermal recrystallization under temperatures up to 900–1100 °C exceeding significantly residual post-shock and causing the partial melting in places. Such high-gradient thermal metamorphism is revealed in impact structures for the first time and thereby paid a special attention. The thermal decomposition of minerals and subsequent blasthesis caused essential changes in magnetic properties of shocked crystalline rocks and produced a high-amplitude magnetic anomaly in the center of the impact structure. In places, specific lithologies denominated as coptoblastoliths formed locally due to high-temperature thermal metamorphism. They are similar by fabric features to "lunar granulites" and give raise a possible origin for the latter. At last, hydrothermal alteration caused by the appearance of the fluid circulation in the heated rocks is widespread. This circulation system may be considered as a new type of hydrogeological convective systems, which was active also during the period of early sedimentation in the crater lake. Due to drilling, both lateral and vertical zonal distribution of alteration minerals is established; it gives an opportunity to elaborate a model of impact-induced hydrothermal system. Based on petrographic evidences, the thermal evolution of the impact crater has been simulated.

The crater fill includes a series of formations: sedimentary variegated mesobreccia; sedimentary megabreccia; polymict mesobreccia; suevites; polymict microbreccia, and tagamite (massive impact melt rocks). The latter either form thin veins within deformed rocks of central uplift (authigenic breccia) or lie within polymict allogenic breccia on its slopes. The suevites occur as irregularly shaped lenses varying sharply in thickness (up to 170 m) within polymict allogenic breccias. The total amount of impact melting products is estimated to not exceed 3.5 km^3. Both tagamites and suevites comprise polycrystalline impact diamonds. The compositional compliance of impactites with target rocks (crystalline with a minor admixture of a sedimentary constituent) is evidenced by various geochemical and isotope data. Higher amounts compared to primary crystalline rocks of some PGE elements such as iridium and osmium may indicate the presence of dispersed matter of the projectile.

Analysis of geological and petrographic information including those gained from the VDW core study combined with numerical modelling allowed a reliable estimate of a number of mechanisms and parameters of the formation of the Puchezh-Katunki impact crater including energy of its formation, which is ca. 1–2 × 10^{28} erg, depth and diameter of the transient cavity, velocity and distance of ejection, duration of individual cratering stages (compression, excavation and early modification etc.). The computer model of the Puchezh-Katunki crater presents a set of variants, simulated the vertical impact of a rocky projectile (asteroid) 3.5–4 km in diameter with the velocity 11–16 km s^{-1}, resulted in a model crater shape, resembling the observed structure geometry. The recorded maximum shock pressures in specimens from the VDW core are close to modeled values, supporting the correctness of both geologic interpretations and general principles of the cratering numerical model.

Thus, Puchezh-Katunki crater is a source of important information, which can answer many questions concerning impact cratering mechanism, behavior of various rock materials under high impulse pressures and temperatures, its movement and further evolution. In this connection, the presented book summarizing the data of many-sided research works in Puchezh-Katunki, can contribute to the development of further studies of impact cratering in general.

The manufacturer's authorised representative in the EU is Springer Nature Customer Service Centre GmbH, Europaplatz 3, 69115 Heidelberg, Germany. If you have any concerns regarding our products, please contact ProductSafety@springernature.com

Printed and bound by CPI Group (UK) Ltd, Croydon, CR0 4YY
30/09/2026
02225602-0004